T0321910

Theory
and Research in
Behavioral Pediatrics
Volume 2

Theory and Research in Behavioral Pediatrics
Volume 2

Edited by

Hiram E. Fitzgerald, Ph.D.

Professor of Psychology
Michigan State University
East Lansing, Michigan

Barry M. Lester, Ph.D.

Director of Developmental Research
Child Development Unit
Children's Hospital Medical Center
Assistant Professor of Pediatrics
Harvard Medical School
Boston, Massachusetts

and

Michael W. Yogman, M.D.

Associate Chief, Division of Child Development
Children's Hospital Medical Center
Assistant Professor of Pediatrics
Harvard Medical School
Boston, Massachusetts

PLENUM PRESS • NEW YORK AND LONDON

ISBN 0-306-41566-6

©1984 Plenum Press, New York
A Division of Plenum Publishing Corporation
233 Spring Street, New York, N.Y. 10013

Printed in the United States of America

Contributors

J. Lawrence Aber, Ph.D. • Department of Psychology, Barnard College, Columbia University, New York, New York

Heidelise Als, Ph.D. • Child Development Unit, Children's Hospital Medical Center and Harvard Medical School, Boston, Massachusetts

Catherine T. Best, Ph.D. • Department of Human Development, Cognition, and Learning, Teachers College/Columbia University, New York, New York

Nancy A. Carlson, Ph.D. • Center for the Study of Youth Development, Stanford University, Stanford, California

Thomas Z. Cassell, Ph.D. • Plymouth Center for Human Development, Northville, Michigan

Dante Cicchetti, Ph.D. • Department of Psychology and Social Relations, William James Hall, Harvard University, Cambridge, Massachusetts

Frank H. Duffy, M. D. • Developmental Neurophysiology Laboratory, Seizure Unit and Department of Neurology, Children's Hospital Medical Center and Harvard Medical School, Boston, Massachusetts

Donna J. Follansbee, M. S. • Department of Psychiatry, Harvard Medical School, Boston, Massachusetts

Stuart T. Hauser, M. D., Ph.D. • Department of Psychiatry, Harvard Medical School, Boston, Massachusetts

Frances Jensen, M. D. • Developmental Neurophysiology Laboratory, Seizure Unit and Department of Neurology, Children's Hospital Medical Center and Harvard Medical School, Boston, Massachusetts

George Mower, Ph.D. • Developmental Neurophysiology Laboratory, Seizure Unit and Department of Neurology, Children's Hospital Medical Center and Harvard Medical School, Boston, Massachusetts

Ross D. Parke, Ph.D. • Department of Psychology, University of Illinois at Urbana-Champaign, Champaign, Illinois

Barbara R. Tinsley, Ph.D. • Department of Psychology, University of Illinois at Urbana-Champaign, Champaign, Illinois

Preface

Volume I of *Theory and Research in Behavioral Pediatrics* focused on issues of early human development, with special emphasis given to assessment of the preterm infant and to factors influencing the organization of the caregiver-infant relationship. Chapters in Volume 2 cover a broader range of topics and encompass a wider age span. Chapter 1 provides a historical review of the relationship between developmental psychology and pediatrics. The authors, Barbara R. Tinsley and Ross D. Parke, discuss differences between behavioral pediatrics and pediatric psychology and note that interdisciplinary collaboration in research and application has increased steadily in recent years. However, if similar collaborative efforts are to occur in education and training of pediatricians and developmental psychologists, it will be necessary to determine just what each discipline hopes to gain from such collaborative efforts. Tinsley and Parke report the results of a national survey designed to determine the areas of developmental psychology that pediatricians perceive to be of potential benefit to them in their delivery of pediatric care. Results of the survey suggest that there are many ways in which developmental psychology could be incorporated into the pediatric curriculum.

In many respects, Chapter 2 sets the stage for the remaining chapters. Nancy A. Carlson and Thomas Z. Cassel argue that behavioral pediatrics research should be guided by a social-ecological theoretical model that is firmly grounded in contemporary evolutionary theory and that stresses the systemic nature of the organism and the environments to which the organism adapts. Their theoretical stance rejects static, linear approaches to bio-behavioral organization because such approaches fail to account for the many transformations that must occur as the individual "adapts to and makes transitions to expanding social-ecological environments." Carlson and Cassel encourage the study of the behaviors and environments of physicians and other members of the health care system in order to determine how they influence patient access to this system, transitions from one component of the system to another, and outcomes of primary care.

The idea of dynamic, evolving, adaptive, developmental systems is implicit in Chapter 3, in which Frank H. Duffy, George Mower, Frances Jensen, and Heidelise Als review evidence for neural plasticity. Drawing on evidence from studies of the visual system of intact and brain-lesioned animals, the authors conclude that brain development is regulated by events occurring in both the organism's internal environment (for example, action of opiate peptides) and by events occurring in the external environment (for example, events that affect the temporal regulation of critical periods). Duffy and colleagues are particularly interested in determining how evidence for neural plasticity derived from animal research can be applied to the study of the premature infant's negotiation of developmental pathways. Their review of the animal literature suggests that "the premature infant's environment may well have a profound effect on subsequent neural development."

In Chapter 4, Catherine T. Best evaluates two prominent theories of infant speech perception. One of these suggests that infants have specialized perceptual mechanisms genetically "tuned" to linguistic contrasts among phonemes. The other view suggests that speech perception is controlled by acoustic properties of the sound stimulus, regardless of whether the auditory stimulus represents speech sounds or nonspeech sounds. Best argues that each of these genetic-based theories are deficient and offers an alternate view that emphasizes the active, information-processing abilities of the infant. This speech source view implies that infants attend to vocal tract information. Moreover, Best suggests that left-hemispheric specialization for language can be linked to the infant's ability to detect information about the articulatory gestures of the speaker's vocal tract. According to this essentially ecological perspective, "auditory perception is the co-detection of the transformations (motions) and structure of the sound source."

In Chapter 5, J. Lawrence Aber and Dante Cicchetti provide a detailed critical review of twenty years of research on the socioemotional development of maltreated children. The authors concentrate on a representative sample of influential studies, using them to illustrate how one's theoretical orientation and research methodology can influence the kinds of conclusions drawn by the investigator. Five types of studies are reviewed: clinical, cross-sectional, follow-up, theoretically derived, and prospective longitudinal. They conclude the chapter by advancing a series of recommendations for improving the quality of research in the area. Among their conclusions is the advocacy of a transactional developmental model, which shares much in common with the socioecological perspective described by Carlson and Cassel.

In the final chapter, Stuart T. Hauser and Donna J. Follansbee examine major theoretical views of the development of ego identity, perhaps the dominant adaptive achievement to be realized during the transition from childhood to adulthood. After a systematic review of the theories of Erikson, Marcia, Hauser, Block, and Loevinger, the authors proceed to an integrative discussion of psychosocial development during adolescence, focusing not only on the growth tasks of this transitional period of human development but also on the role of the family in the transitional process. They suggest that lack of interdisciplinary collaboration impedes development of a comprehensive "understanding of the contextual aspects of adolescent development."

Catchwords for the offerings in Volume 2 include: system, organization, context, ecology, interdisciplinary collaboration, and transition. Collectively, these chapters suggest that behavioral pediatric research may be a medium in which the various social, behavioral, and life sciences can coalesce, yielding an integrative understanding of human development and providing a base for comprehensive health care.

HIRAM E. FITZGERALD
BARRY M. LESTER
MICHAEL W. YOGMAN

Contents

The Historical and Contemporary Relationship between Developmental Psychology and Pediatrics

A Review and an Empirical Survey

BARBARA R. TINSLEY AND ROSS D. PARKE

Keep drugs and their properties in memory, but remember, it is in the mind where you will find the cures of disease.

—Hippocrates

Developmental psychologists and pediatricians share a common goal: facilitation of optimal growth and development in children. The relationship between the study of child development and pediatric medicine would appear to hold considerable interest to academics and those with applied interests in both disciplines. Yet, until recently, the members of each profession have been reluctant to collaborate in research, teaching, or clinical work. It is possible to hypothesize and historically trace three main reasons for this apparent indifference of developmental psychologists and pediatricians regarding each other's discipline.

Barbara R. Tinsley and Ross D. Parke • Department of Psychology, University of Illinois at Urbana-Champaign, Champaign, Illinois 61801. Preparation of this chapter was supported in part by NICHD Grant HEW PHS 05951, NICH Training Grant 7205-01, and a grant from the National Foundation March of Dimes. The chapter was completed during Ross D. Parke's tenure as a Belding Scholar at the Foundation for Child Development.

First, both physicians and psychologists have viewed their disciplines as separate and in no need of supplementation from other disciplines. Although physicians were among the first to focus attention on prenatal care and infant mortality long before World War II, they have been slower to focus on developmental concerns because child development has often been viewed by the medical establishment as outside its purview (Senn, 1975). Similarly, psychologists have been slow to recognize the research and application possibilities of psychological factors within the area of health care (American Psychological Association Task Force on Health Research, 1976).

Another part of the reason for this separation stems from fundamental differences in the types of models that have guided research in pediatrics and child development. On the pediatric side, the traditional and dominant medical model of disease, which conceptualizes disease as a deviation from normative biological functioning, does not address social, psychological, and behavioral dimensions of disease (Engel, 1977, 1980). In contrast, psychologists have concentrated on behavioral models that give little explicit recognition to biological factors. These models have dominated applied and theoretical child development research approaches for many years. Acceptance of medical models leaves the physician without perceived need for the skills and knowledge of trained psychologists, and reliance on behavioral models, in turn, has discouraged psychologists from focusing on research and training in the biological basis of behavior and on preventative and rehabilitative health care.

Third, as a result of these differences in models, a historical conceptual confusion exists concerning how to define the value and implementation of the pediatric-psychological collaboration. Until very recently, models have been unavailable to guide the development of pediatric-psychological collaboration.

In this chapter, these issues that have accounted for the lack of a relationship between developmental psychology and pediatrics will be examined in greater detail as well as the recent shifts in the nature of the models of both pediatric and developmental psychology that are providing a stronger basis for linkages between these two disciplines. Finally, a new survey of pediatricians that provides guidelines concerning the areas of developmental psychology that may be of most value to pediatrics will be presented.

1. THE HISTORY OF THE RELATIONSHIP

The history of the relationship between pediatrics and child development probably began prior to the turn of the century with the publication of clinical psychologist Lightner Witmer's "The Common Interests of Child

Psychology and Pediatrics" in the journal *Pediatrics* in 1896. Although interest in child developmental issues was expressed by a few highly visible pediatricians during the early decades of the twentieth century (e.g., Arnold Gesell, Alfred Washburn, and C. Anderson Aldrich), research and teaching in the area of child development have not been a major concern of either academic or clinical pediatrics (Richmond, 1967). This lack of pediatric concern and knowledge regarding developmental issues has existed through the years despite the fact that a functional knowledge of normative development would appear to be a necessity for maintaining child health. Although it is true that physicians have sporadically voiced interest in such infant care practices as feeding schedules, weaning, swaddling, and so forth, concerted pediatric involvement with issues of child development has been absent or at least not formally articulated (e.g., journals and books).

Pediatricians who have been very interested in normative development and behavioral issues have been largely relegated to functioning as child-rearing educators of the lay public. Suggestions by well-known pediatricians, such as Benjamin Spock or T. Berry Brazelton, are more extensively followed by parents than by academic pediatricians.

In an oral history of the child development movement in the United States beginning in 1920, Senn (1975) reflected on pediatric interest in child development:

> Pragmatists, by and large, both pediatrician clinicians and researchers have traditionally thought in terms of immediate solutions to problems, of the prevention and cure of specific illnesses, and have been included to shrug off theories of development and behavior as outside their spheres of concern. (p. 56)

Even though such current points of contact between pediatrics and developmental psychologists as follow-up studies of high-risk infants and children (preterms, physically handicapped, and mentally retarded) can be traced back to the 1920s, these studies were often implemented by physicians, psychiatrists, or educators and not child psychologists (Kopp, 1983). By the early 1930s, a certifying board was established to set standards for subspecialty training in pediatrics (Yogman, 1980). In the 1940s and 1950s, the field of pediatrics began to shift in a variety of ways that laid the groundwork for more liaisons with other disciplines, especially developmental psychology. As Kopp (1983) notes, during this period, "the specialty of pediatrics began to develop into a distinct medical group with criteria established for training and board certification" (Kopp, 1983, p. 1083). There was increased attention to the perinatal problems and postnatal mortality and morbidity rates. Specialized centers for neonatal care were built after World War II accompanied by advances in technology that, in turn, increased the research activities of pediatricians. At the same time, environmental factors both during pregnancy and in the postpartum period were beginning to be recognized as being of potential value for understanding later child outcomes. These shifts, in part, laid the groundwork for stronger ties between pediatrics and developmental psychology.

Beginning in 1950, the field of pediatrics seemed to become more sensitive to child development issues. In his presidential address to the American Pediatric Society (1950), S. Z. Levine noted the necessity of increasing emphasis on all aspects of the child, in contrast to the narrow concern with biological issues. He suggested that pediatricians must also be concerned with children's environments, child-rearing practices, and the consequences of these for the child's development.

In a landmark editorial in 1951, the American Board of Pediatrics issued a statement concerning the importance of training in child growth and development for pediatricians (Harper, 1951). Citing a 1950 study by the American Academy of Pediatrics, which found that on an average day pediatricians spend 54% of their time in the health supervision of well children, the editor of *Pediatrics* (journal of the American Academy of Pediatrics), called for more stringent pediatric examination requirements with regard to growth and development that would require not only memorization of facts, but an understanding of how to implement developmental information in daily pediatric practice. Harper stated:

> The concept of growth and development should remain a broad one. . . . We are primarily interested in providing the prospective pediatrician with the tools for the most optimal practice. This means the teaching of physical, mental, emotional, and social growth and their interrelationship since one of the most important functions of the pediatrician is the guidance of the total growth, development, and ultimately social adjustment of the individual. (p. 431)

In a later issue of *Pediatrics* in that same year, Arnold Gesell (1951) firmly endorsed the concept of incorporating child development into pediatric training. The trends continued and formal recognition of the importance of child development for pediatrics came in 1960 when the American Academy of Pediatrics (AAP) established the Section on Child Development (American Academy of Pediatrics, 1960). By this action, the AAP declared child development "a basic science for pediatricians" (Richmond, 1967).

Nor was this interest a one-sided affair. Psychologists, in turn, were recognizing the implications of the findings of child development research for pediatrics. In 1965, developmental psychologist Jerome Kagan published an extremely influential statement concerning the necessity for pediatric concern with developmental issues entitled "The New Marriage: Pediatrics and Psychology." Kagan suggested that the pediatrician who had knowledge of developmental research strategies and results would "have the opportunity to think theoretically about personality dynamics—about the etiology of symptoms—and acquire the mental set that permits subordinating a concern with the manifest appearance of a symptom to a concern with the dynamics of symptom growth and maintenance" (p. 272). Similarly, in the late 1950s and early 1960s, a small group of developmental researchers began to investigate the implication of prematurity and other birth-related trauma for later development, which, in turn, resulted in early collaborative projects involving pediatric and child development researchers (e.g., Graham, Pennoyer, Caldwell, Greenman, & Hartman, 1957; Werner, Simonian, Bierman, & French, 1967). This type of research effort indicates an awareness that the influence between the disciplines of pediatrics and child development is bidirectional, and that both disciplines as well as children's welfare benefit from collaboration of professionals in both areas.

On the pediatric side, two influential medical researchers, Julius Richmond and Leon Eisenberg, continued the plea for collaboration in terms of both research and physician training. Richmond (1967) suggested that, given the need to understand children in more dynamic terms, pediatricians occupy a perfect position in which to observe children's development and maturation. This might best be accomplished, suggests Richmond, by pediatric research in developmental areas. At about this time, pediatrician Richmond was beginning a series of collaborative studies with psychologists such as Betty Caldwell when both were at Syracuse University. Similarly, the Perinatal Collaborative Study, which began in 1959, was further evidence of the recognition of the importance of development as well as a further sign of working relationships between pediatrics and developmental psychology.

This trend toward emphasis on developmental training for pediatricians continued to be endorsed by others. Eisenberg (1967) stated that

> pediatrics has long neglected its responsibility for the study of child development. The central nervous system, which quadruples its mass and undergoes an explosive proliferation in its interconnections as the infant progresses from birth to adolescence, is the most distinctive feature of the human species; it is central to the understanding of the integrated function of the developing child. . . . American pediatrics . . . has moved toward an ever more precarious preoccupation with the exotic; it prepares its house officers exceedingly well for what they will rarely see and says hardly a word about what will confront them daily. (p. 656)

The 1970s was a period in which there were increasing numbers of collaborative arrangements for both training and research between pediatricians and psychologists. Collaborative research arrangements between pediatricians and psychologists have yielded valuable information concerning the cognitive outcome of premature infants, expanding our understanding of the importance of very early parent-infant interaction, and suggested new strategies for the development of infant evaluative measures (e.g., Brazelton, 1973; Field, Sostek, Goldberg, & Shuman, 1979; Kearsley & Sigel, 1979; Littman & Parmelee, 1978; Parmelee, Kopp, & Sigman, 1976).

Increased lay awareness of the need for child care and child health services has been the impetus for such social service programs as Head Start, Maternal and Infant Care, and Children and Youth Programs. The 1970 White House Conference on Children and the designation of 1979 as the International Year of the Child are other examples of recent public interest in child welfare. These renewed commitments to child services have brought new pressure for pediatricians to be sophisticated in the area of child development (Richmond, 1975):

> From a perspective of 30 years of attempting to teach child development in pediatrics, it is unfortunately necessary to record that we still have a long way to go to realize the objective that the American Board of Pediatrics expounded in 1951: ". . . that pediatric training centers must increasingly assume the responsibility for the day-to-day teaching of growth and development in a clinical setting. This means a practical evaluation of the total development of each patient and its meaning to parents and physicians." (p. 430)

However, Richmond remains optimistic that child development has a firm position as a discipline for pediatricians. Recently, Richmond and Janis (1980) stressed the need for an interdisciplinary focus in understanding children's development that includes biological, psychological, and social factors.

Pediatric departments are being actively encouraged by the American Academy of Pediatrics to develop curricula more pertinent to the needs of children within modern pediatrics (Kempe, 1978). Green (1982) reports that residency programs accredited by the Residency Review Committee for Pediatrics are required to include behavioral and developmental information. In addition, the American Academy of Pediatrics' Committee on the Psychosocial Aspects of Child and Family Health prepares psychosocial content materials for anticipatory guidance from the prenatal period through adolescence. Furthermore, the federal government has provided funding for such programs under the Health Manpower Act of 1976 (Davis, Stone, Levine, & Stolzenberg, 1980), and the National Institute of Mental Health has increased its support of cooperative teaching and consultation by primary health care providers and psychologists (Tuma & Schwartz, 1978).

Lastly, programs in behavioral pediatrics are being developed, such as those offered by the William T. Grant Foundation and by the General Pediatrics Academic Development Program of the Robert Wood Johnson Foundation (Green, 1982; Haggerty, 1979).

To fully understand the reasons for pediatrics' shift toward child development, we turn to an examination of the medical model, the recent changes in the model, and the implications of this shift for the collaborative issue.

At the same time, the changes within developmental psychology toward an increasing recognition of biological influences on development will be noted. In part, this will provide a better basis for understanding patterns of increasing collaboration that have emerged in recent years.

2. THE MEDICAL MODEL AND PEDIATRICS-DEVELOPMENTAL PSYCHOLOGY COLLABORATION

The dominant medical model of disease is biomedical, emphasizing molecular biology as its basic discipline. This model assumes disease to be fully explained by biological variables, disregarding social, psychological, and behavioral dimensions of illness (Engel, 1977). To the extent that this model is strictly adhered to, collaboration with other disciplines is unnecessary for fulfilling the basic aims of pediatrics. However, it is becoming increasingly recognized by the medical establishment that this concentration on biomedical factors of disease and the resultant exclusion of psychosocial principles interferes with optimal diagnosis and management of disease.

Engel (1977) has suggested that the presence of a biomedical defect at best presents a necessary but not sufficient condition for occurrence of disease. Active disease, according to Engel, is more accurately conceptualized as the complex interaction of the patient's biochemical defect and other cultural, social, and psychological factors. Engel calls for a "biopsychosocial model" of disease treatment and health care:

> Medicine's unrest derives from a growing awareness among many physicians of the contradiction between the excellence of their biomedical background on the one hand and the weakness of their qualifications in certain attributes essential for good patient care on the other. Many recognize that these cannot be improved by working within the biomedical world alone. (p. 134)

Similarly, patient care is very dependent on the existing model of disease. It is suggested that medical outcomes are often inadequate not because of a lack of appropriate technology but because conceptualizing about disease and its etiology is inadequate.

Others have focused on the ways in which this biomedical model of disease has structured medical education. In directing attention to the proliferation of medical specialties, Pattishall (1973) suggests that medical training programs have emphasized science, technology, and specialization at the expense of humanistic, integrative medicine. This practice results in an exclusive focus on the disease rather than a concern for the patient. He states that "the physician-in-training receives very little exposure to the behavioral factors involved in illness or health" (p. 923).

The dominant medical model of disease has deemphasized normal development to the extent that the main prior link between medicine and psychology has been psychiatry, which as a discipline is organized around abnormality. Wexler (1976) suggests that medical study of human behavior begins with psychiatry. However, child psychiatry differs substantially from behavioral pediatrics, although there is some overlap. The traditional focus of psychiatry, according to Friedman (1975), has been treatment based on psychoanalytic concepts and techniques, whereas behavioral pediatrics is eclectic, with major reliance on the discipline of child development. Friedman states, "though many child psychiatry programs have active pediatric liaison activities, the field has not, in general, been well integrated into major health delivery systems for children" (p. 515).

Richmond (1975) portrays the child psychiatrist as a referral resource for the pediatrician because the psychiatrist is more theoretically oriented and deals routinely with severe behavioral and psychological problems. Richmond characterizes the pediatrician as more concerned with "relief of symptoms" rather than longer term treatment plans because of practice constraints. Although the pediatrician is acquiring competence in psychological assessment and other developmental areas, Richmond suggests that primary prevention and detection of psychological disturbance must eventually become the responsibility of pediatricians since they are often the first and most accessible contact between the disturbed child and mental health services.

Evidence of dissatisfaction with the medical model has come from psychologists as well as physicians. The American Psychological Association Task Force on Health Research (1976) concurred in these opinions concerning the relationship between the biomedical model of disease and the contribution of psychology to medicine. The task force suggested that the etiology of disease is multifactorial, dependent on the interaction of the virulence of infection and the susceptibility of the host. In turn, the host's susceptibility depends on the individual's predisposition to succumb selectively to various assaults on its physical integrity as a function of early life experiences, economic status, and social status.

Similarly, Logan Wright, a psychologist, in his presidential address to the Southwestern Psychological Association suggested that psychologists' "disdain" for the medical model may have made viewing the role of psychology in treating physically ill people seem unimportant. Psychologists are discouraged from focusing on social or psychological environmental factors affecting disease because the prevailing medical model suggests that physical illness derives only from biophysical conditions and, at the same time, does not acknowledge the relationship between the effects of disease and the psychological symptoms and manifestations of physical disease. Wright (1976) states:

My judgment is that over one half of all the psychopathology of children in this world is related to a physical behavioral symptom of some sort for which help is usually sought first from a dentist or medical doctor; and that *less than* one half of the psychopathology of children falls within the traditional psychological categories of neurosis, psychosis, and character disorder. (p. 16)

Interest in primary care and family medicine may be viewed as evidence of renewed commitment to a physician approach that incorporates psychosocial issues (Engel, 1977; Hansen & Aradine, 1974; Richmond, 1975). The trend toward primary care in medicine emphasizes the broader definition of health care: delivery of comprehensive health services to a population. Rather than focusing on disease management, physicians who identify themselves with primary care consider their task to be defining, acting on, and caring for all health related problems (Davis, Stone, Levine, & Stolzenberg, 1980). According to a recent survey by Burnett and Bell (1978), 80% of the respondent physicians described their practices as the provision of primary health care. Hansen and Aradine (1974) suggest that the broad definition of health associated with pediatric primary care includes social, emotional, and behavioral dimensions, therefore expanding "the expectations which society has placed upon pediatrics" (p. 246).

Richmond (1967) summarizes the major issues concerning the new medical model that may be evolving:

the concept of human development must become a central theme of medical education if it is to meet the changing needs of society. I again come back to a shift in human biology—of which medicine is a division—from an emphasis on disease to one of process. And processes may be ordered around development as a unifying concept. (p. 656)

The dominance of the strict biomedical model of disease is lessening, and a model incorporating concern with behavior and the "whole" person is much more prevalent in research, academic, and clinical settings. Wexler (1976) observes that the medical school is becoming an important intellectual center for ideas and research regarding psychosocial issues and their effect on health and disease. Moreover, pediatrics has been observed to offer a unique opportunity for systematic study of developmental research questions. Suggesting that the research potential associated with pediatric-psychology liaisons is its "most exciting facet," Tuma (1975) proposes that by studying the effects of illness on children we may gain understanding concerning the process of adjustment to stress in children. Some recent collaborative work between developmental psychologists and pediatricians on the impact of childhood illness on adult behavior (Haskins, Hirschbiel, Collier, Sanyal, & Finkelstein, 1981), and infant cognitive functioning (Haskins, Collier, Ramey, & Hirschbiel, 1978) have recently appeared.

3. THE CHILD DEVELOPMENT APPROACH AND PEDIATRICS-DEVELOPMENTAL PSYCHOLOGY COLLABORATION

Developmental models have undergone a series of revisions as well over the past two decades that have made collaboration between pediatrics and child development more likely. Increased recognition has been given to biological factors in development in terms of constitutionally based differences among infants and children in a variety of temperamental characteristics (Thomas, Chess, Birch, Hertzig, & Korn, 1963). Others argue that infants are biologically "prepared" for certain kinds of learning (Hinde & Stevenson-Hinde, 1973; Sameroff, 1971) or for particular social experiences (Schaffer, 1971, 1977). Nor is the recognition of biological factors restricted to genetic or constitutionally based differences. The continuing importance of biological factors, such as adequate nutrition, for proper cognitive social development is increasingly recognized (Ricciuti, 1980).

At the same time, however, new developmental models that incorporate both biological and experiential components have appeared. These models demonstrated that neither biological nor experiential factors alone yield adequate understanding of development and only by combining these components can one better understand development. Sameroff's (1975) transactional model of development suggests that two continua—a continuum of reproductive casualty, which includes both genetic constitutional factors as well as birth-related trauma, and a continuum of caretaking casualty, which includes the social, intellectual, and physical environment—are necessary for adequate prediction of developmental outcomes. Other recent models such as systems theory approaches similarly stress the necessity of considering the complex interplay between biological and experiential factors (Ramey, MacPhee, & Yeates, 1982; see Sameroff, 1982).

Much of the current collaborative research effort between pediatricians and psychologists is concerned with infancy. Such main collaborative projects as Parmelee's investigations of the effects of prematurity and Brazelton's efforts to incorporate social items in measures of infant competence suggest that, as pediatricians and psychologists move toward each other's theoretical models and perspectives, important collaboration in such areas as research in infant development is emerging. In fact, many of the recent collaborative efforts explicitly recognize the transactional model as the theoretical framework for their research (Sigman & Parmelee, 1979; Zeskind & Ramey, 1981).

However, some collaborative arrangements have a different history, namely a shared concern with treatment of childhood behavior problems, such as childhood fears (Melamed, 1980) and developmental disabilities (Shonkoff, Dworkin, Leviton, & Levine, 1979). Some of these links are due to advances in behavior modification strategies. (See Melamed, 1980, for a review of this work.)

By the mid 1970s, there was considerable evidence that the traditional medical model was undergoing modification and that developmental psychology was paying increased attention to biological factors in development. These dual sets of changes, in turn, increased the likelihood of collaborative arrangements between pediatricians and psychologists.

4. REDEFINITION OF THE ROLE OF THE PEDIATRICIAN

Because of a variety of factors, including demographic shifts, secular trends, changes in mortality rates, and modification of the consumer's role in defining medical practice, the role of the pediatrician is undergoing a series of changes, which, in turn, make the psychologist-pediatrician collaboration more likely (Green, 1982; Tuma, 1975; Yogman, 1980).

Owing to the greater mobility of families, parents are less able to rely on the availability of the extended family for child care information. Thus, new parents are relying more heavily on professionals, including pediatricians, for this type of information. Second, because of a decrease in child mortality and morbidity rates, at least partially as the result of advances in preventative medicine, pediatricians find less of their time involved with sick children, and are more often being asked by parents how to deal with well, active children (Rogers, Blendon, & Hearn, 1981). Parent requests for child-rearing advice are taking increasing amounts of the pediatricians' professional time. Estimates of the extent to which pediatric primary care visits involve child-rearing, behavior problems, or other psychological components vary from 37% to over 50% (Duff, Rowe, & Anderson, 1973; McClelland, Staples, Weisberg, & Bergen, 1973).

Although it was once hypothesized that pediatricians' lack of concern with behavioral issues is due to parents' disinterest in their children's behavioral and social/emotional development, this has not been shown to be the case. Chamberlin (1974), in a longitudinal survey and interview study of parents of 200 children from ages 2 through 5, found that parents were very concerned with their children's behavioral and emotional development. Most of the mothers who were concerned about their child's behavior had not talked with a professional about their concern, including the child's pediatrician. However, mothers who did consult their pediatricians about their child's behavioral or emotional problems "generally found the discussions to be worthwhile." Hornberger, Bowman, Greenblatt et al (1960), in a survey of 800 California mothers, found that 91% of these mothers indicated at least some mild concern about some aspect of their child's behavior and 28%

suggested that their child's behavior was causing serious alarm. Finally, in a routine screening program for indigent school-aged children, mothers spontaneously expressed worry about some aspect of the child's health in 155 of 303 cases (51%). For these 155 children, 48 were behavioral complaints (Starfield & Borkowf, 1969). There is little support for the parental disinterest hypothesis as a reason for the lack of pediatric interest in developmental psychological issues. Third, as discussed earlier, emphasis in medical school training is shifting from a focus on the traditional biomedical model of disease diagnosis and management to treatment of the "whole person," the result of which is increasing importance of the behavioral sciences in medical training (e.g., developmental psychology training for pediatricians).

It is apparent that pediatric practice is at least partially shifting from a principal concern with disease and morbidity to inclusion of services addressing behavioral and developmental issues. Consumers not only seek specific behavioral and developmental information from pediatricians but also expect them to act as competent resources for guidance concerning normal development and child rearing practices (Hansen & Aradine, 1974). However, comparative study of parents' and pediatric residents' knowledge of normal and abnormal development (Shea & Fowler, 1983) suggests that, although the pediatric residents were significantly more knowledgeable than parents, the physicians still harbored basic misconceptions in these areas. Moreover, the physicians indicated feeling more comfortable discussing traditional medical problems than behavioral or developmental issues with parents.

Recently, there has been an effort to formalize a link between pediatrics and developmental psychology by the introduction of a new pediatric area of special interest: *behavioral pediatrics.*

4.1. Behavioral Pediatrics

Fitzgerald, Lester, and Yogman (1981) recently defined the task of this new area.

> The task of behavioral pediatrics as a newly emerging discipline is to develop a basic science that links biological science with behavioral science to provide a clinically relevant perspective on early human development. (It is unnecessary to restrict the goals to "early" development, however, and the task should cover the full range of development.) Behavioral pediatrics may be considered a subspecialty of the respective fields of developmental psychology and pediatrics. In developmental psychology this would be akin to education, cognition, or physiology and related to but different from pediatric psychology, which emphasizes the clinical case of children in psychiatric settings. In pediatrics, this would parallel subspecialties such as cardiology, or neurology, particularly with respect to research and training activities. (p. 3)

Many of the collaborative projects that are noted in our review would fall within this definition of behavioral pediatrics. Training in behavioral pediatrics is generally at the postdoctoral level for psychologists and post-residency training programs for physicians. A number of programs that offer interdisciplinary research and training in child development and pediatrics have begun to appear in the past few years (see Haggerty, 1979) and are established generally in a medical setting.

The areas in which developmental psychology can contribute to the discipline of pediatrics will be considered next.

5. DEFINING PSYCHOLOGY'S ROLE IN PEDIATRICS

As discussed thus far, behavioral science is still not uniformly accepted by the medical community (Wexler, 1976). One possible explanation for this is that behavioral scientists have not effectively translated their theories, methods, or knowledge base into a format that has clinical applicability. Physicians, because of time constraints in both training and practice, demand specific skills with which they can solve specific problems. Therefore, the implication is that, until psychology can provide practical skills that can be applied to a variety of practice-related problems, physicians will continue to be skeptical about the value of the psychologist's contribution to pediatrics.

Furthermore, in pediatrics, confusion remains as to exactly what it is that developmental psychology has to offer. Work (1955) suggests that there is an "apparent paradox between knowledge gained [in child psychology] and the knowledge which others (pediatricians) had lumped together in the words 'common sense' " (p. 16).

In what specific areas can child development and developmental psychology contribute to academic and clinical pediatrics? Kagan (1965) suggests that psychology offers pediatricians the opportunity to better understand the relationships between prenatal and perinatal casualty and later behavior problems. Furthermore, according to Kagan, pediatrics needs assistance in the early detection of severe childhood disturbance, delinquency, and other psychopathology. Kagan lists a series of six tasks for pediatrics that can best be accomplished with the help of psychologists: (a) development of better assessments for infants-at-risk (e.g., habituation and memory sensory tasks); (b) agreement on a concept of intelligence; (c) development of a measure of the rate of intellectual growth; (d) better understanding of the child effect of mother disturbance (to determine biological vs. environmental causation of psychopathology); (e) improved pediatrician-parent relationships; and (f) training of the pediatrician to understand personality in the same way he or she understands disease.

As Kagan (1980) has noted in a fairly recent update of the contribution of developmental psychology to pediatrics, considerable progress has been made in these areas since 1965. In addition, other areas in child development relevant to pediatric practice have also been explored in recent years, including the development and treatment of hyperactivity (Werry & Sprague, 1970; Whalen & Henker, 1976), child abuse (Helfer & Kempe, 1974), eating disturbances (Bemis, 1978; Siegel & Richards, 1978), reading problems (Benton & Pearl, 1978), the effects of child care arrangements (Belsky & Steinberg, 1978; Kagan, Kearsley, & Zelazo, 1978), the development of the attachment relationship (Ainsworth, 1973; Sroufe & Waters, 1977), and the role of the father (Lamb, 1976; Parke, 1981) and peers (Asher & Gottman, 1981; Lewis & Rosenblum, 1975) in social development.

In a review of child development studies that have important implications for pediatric practice, Chamberlin (1973) suggests that knowledge and research skills within the discipline of child development are necessary for academic and clinical pediatricians. Specifically, Chamberlin cites research in social development (e.g., child temperament studies, behavior modification, family interviewing strategies, and the effect of early perinatal mother-child contact) and cognitive development (e.g., effects of home environment on intellectual functioning) as important for pediatricians to acknowledge. Although knowledge gained from individual research efforts in the area of child development obviously is important for pediatric training and practice, it has been suggested that the most valuable contribution of psychology to medicine is its methodology for asking and seeking answers to questions (Dacey & Wintrob, 1973). In addition to providing physicians with appropriate knowledge concerning human behavior, the discipline of psychology also offers physicians alternative patterns of conceptualization concerning clinical diagnosis and management of disease. Physicians are offered the opportunity to interpret patient data theoretically and with a process orientation.

Psychology has much to offer clinical and academic pediatrics, ranging from very specific behavioral strategies for clinical pediatrics to proven research methodologies for academic pediatrics.

5.1. Pediatric Psychology

The interface between developmental psychology and pediatrics has yielded a new but well-defined subspecialty incorporating both areas: *pediatric psychology*. The pediatric psychologist is trained in both child development and child clinical psychology and contributes to patient management with psychological diagnosis and treatment skills.

Logan Wright (1967) extends the definition of a pediatric psychologist to include a philosophical distinction. Ideally, according to Wright, the pediatric psychologist should "maintain a conviction that the behavioral sciences have uncovered certain principles of human behavior which can benefit parents and seek to generate new knowledge regarding such principles as well as ways of communicating these principles to parents" (pp. 323–324).

The pediatric psychologist differs from the clinical child psychologist by practicing in a pediatric setting rather than in a psychiatrically oriented setting and because the target population of pediatric psychologists is children with behavioral disorders precipitated by disease or disease treatment rather than the clinical child psychologist's population of behaviorally deviant children. Unfortunately, the profession of pediatric psychology has as its focus clinical assessment and intervention and has not conducted collaborative research with pediatricians.

Training of pediatric psychologists usually consists of practicum experience both before and after obtaining a degree in developmental or clinical child psychology (Routh, 1980). This practicum training is most often offered in medical school departments of pediatrics or child development clinics, usually at the internship level, but also in the form of postdoctoral appointments. Actual training emphasizes infant evaluation and pathology, perceptual evaluation, programs for handicapped children, diagnosis and psychotherapy for developmentally deviant children, and nursery school consultation and programming (Routh, 1970, 1972, 1980; Tuma, 1975; Wright, 1967). Relatively less emphasis is placed on research experience during training. Although several models exist for the physician-pediatric-psychologist liaison (Stabler, 1979), the most common model is consultation after referral. Tuma (1975) suggests that pediatric psychologists can best be used to

> help pediatricians understand normal psychological growth and development; help counsel parents about childrearing issues; to recognize, and where possible, remediate disorders of development. Diagnostic services of the pediatric psychologist also involve development of intellectual appraisal determined by specialized techniques not usually found in the clinical child psychologist's practice. (p. 11)

However, there are a variety of models that can guide pediatrician-psychologist collaboration; behavioral pediatrics and pediatric psychology are only two models. Another model assumes that, in the near future, only a small proportion of pediatricians will have direct access to a psychologist in their routine practice. Therefore, increased emphasis on developmental psychological issues in medical education may be a more practical approach.

6. INCREASING PEDIATRICIAN'S KNOWLEDGE OF DEVELOPMENTAL PSYCHOLOGY

6.1. Current Status

Given the apparent increasing interest of pediatrics in understanding development, how active are psychologists in medical education? In 1949, Page and Passey, in a survey of the 70 medical schools in the United States, found that only 40% offered psychology courses to their students. A series of surveys of the professional activities of psychologists in medical schools (Matarazzo & Daniel, 1957; Nathan, Lubin, Matarazzo, & Persely, 1979; Wagner, 1968) found encouraging trends over time: (a) increasing behavioral orientations in medical training and (b) impressive growth in the number of psychologists in medical schools. However, the frequency of collaborative research between psychologists and physicians remained stable. This suggests that *although psychologists are being increasingly employed by medical schools, interdisciplinary research is not the foregone outcome.* Indeed, the respondents reported one-third of their time was spent in research and less than 15% in formal teaching. Even more revealing was the finding that pediatric departments accounted for only 6% of the departments in which the respondents held appointments.

The extent to which psychologists are involved in training pediatricians was explored by Routh (1970, 1972). In 1969 and 1971, he surveyed the one hundred major medical schools in the United States concerning the training opportunities available to medical students. The survey results indicated developmental psychology opportunities existed in several medical schools in 1969 and 1971. In 1969, 65 pediatric departments were involved in some sort of "psychological activity," and this number had increased to 73 departments by 1971. Routh (1972) concluded, "training activities of an applied kind have increased at almost every level of training, and psychological research is at least holding its own" (p. 588).

It seems apparent that psychology is being represented in medical school staffs to an increasing extent. However, is it enough? Several authorities are not convinced that we have achieved an adequate level of either behavioral research or training in medical schools.

In his presidential address to the Society for Pediatric Research (SPR), Wallace W. McCrory presented data from a 1965 SPR questionnaire study of medical school pediatric departments in the United States. Respondents reported that, of 811 ongoing research projects within the 75 pediatric departments, only 32 were related to behavioral science. Furthermore, thirty years of SPR meeting programs showed less than 5% of the delivered papers were reporting about behavior or personality development research (Richmond,

1967). Training in behavioral aspects of development is also limited. Castelnuovo-Tedesco (1967) evaluated the attitudes of American and Canadian physicians toward their psychological training. Results indicated that 49% of the respondents felt this training was inadequate preparation for the demands of their medical practice. A report on the future of pediatric education by a special task force of the American Academy of Pediatrics under the leadership of C. Henry Kempe suggests that 54% of pediatricians who recently completed residency programs felt that their training had been inadequate in biosocial and behavioral areas (Haggerty, 1979).

Thus, although there appears to be increasing concern for behavioral training for physicians, this concern has not yet been satisfactorily reflected in medical school curricula and research efforts.

6.2. Assessment of the Perceived Need for Developmental Psychology Information in Pediatric Training Programs

Empirical assessment of pediatric interest in developmental areas has only recently been the focus of research efforts. However, early pioneering data addressing this question were derived from a study in which this issue was only incidentally explored. In 1950, Blum conducted an observational and interview study in an effort to document the extent to which pediatricians function as psychotherapeutic agents for mothers during well-baby visits. Children were from 3 weeks to 2 years of age. She obtained data concerning the range of topics initiated during the children's examination by both pediatricians and mothers concerning various facets of child care. The mothers' most frequent general concerns were feeding (30%), physical disorders (22%), and aspects of development (14%). Specific concerns of mothers were hunger, spitting, and vomiting. Differences with regard to concerns and age of child were significant. Feeding, undesirable habits (especially thumbsucking), and general development were of most concern at ages 0–2 months, 3–8 months, and 9–26 months, respectively.

More recently, the survey approach has been utilized in order to directly assess pediatrician interest in developmental issues. However, these survey efforts have been limited because of sampling restrictions and a narrow range of topic areas. Toister and Worley (1976) evaluated pediatrician concerns in a single region, South Florida. A 3-page, 16-item questionnaire was administered to 100 pediatricians in that area. Items included soliciting training recommendations for physicians in behavior management and development, evaluation of such training for practice, and the most frequent behavioral questions asked by parents (0–2 years, 2–5 years, 5–7 years, 7–11 years, and 11–17 years). Of the 61 respondents, 80% said more than 10% of daily office visits and phone calls involved "specific requests for behavioral information and guidance." All respondents considered formal training in behavior and development as important for pediatricians-in training. Of 45 topics listed by these pediatricians as problematic, bladder and bowel control (across all ages) and discipline (2–11 years) were cited most frequently, followed by feeding or eating problems (birth–5 years), school fears (5–7 years), and school performance (7–11 years).

Schroeder (1979) documented the concerns of parents responding to child development counseling and behavior management advice provided over the telephone, in brief office visits, and in group evening meetings by a group of pediatricians in a single private practice. The most frequent areas of parent concern were negative behavior, toilet behavior, developmental delays, school problems, sleeping problems, personality problems, sibling/peer problems, and divorce or separation. The generalizability of these findings is restricted because they came from only one pediatric practice.

Finally, in a survey of pediatricians in New England, Dworkin, Shonkoff, Leviton, and Levine (1979) found that 79% of the 97 respondents viewed their training in developmental issues as inadequate. Almost 50% of the sample viewed medical school as being of no value as a source of knowledge in developmental pediatrics. Instead, most (86%) viewed their clinical experiences and current interdisciplinary contacts with neurologists, nurses, social workers, psychologists, and other professionals as more important sources of developmental information. However, pediatricians did not view such practical experience as an adequate substitute for formal training in this area.

The available literature measuring pediatricians' interest in developmental knowledge is an inadequate basis for guiding development of a child-oriented educational program for pediatricians. Further assessment of pediatricians' interest that overcomes the limitations of earlier research efforts is clearly necessary.

7. THE CURRENT SURVEY

A new survey, that aims to provide a sounder basis for design of a pediatric curriculum that involves child development information was conducted. The aim was to provide a more adequate assessment of the pediatrician's perceived need for developmental information. A national survey of pediatricians was completed to accomplish this aim.

7.1. Method

7.1.1. Sample

Under the guidance of the University of Illinois Survey Research Laboratory, a national sample of 190 pediatricians in the United States was randomly selected from the Directory of the American Academy of Pediatrics. The return rate for the sample was 116 or 61% of the total sample (including the original mailing and two additional repeat mailings). The average age of the respondents was 44.2 years; 85% of the sample were male and 15% were female. Seventy-four % had taught in a medical school at some point in their professional career and 53% were currently teaching in a medical school.

7.1.2. The Survey

A 145-item survey consisting of the following general categories was developed: behavior problems, developmental testing, childhood competencies, disabilities, language development, retardation, nutrition, obesity, child abuse, and death and dying. For behavior problems, approximately 12 subtopics such as hyperactivity, aggression, discipline, and fears were included. These topics were represented at 5 age groups: birth–2 years, 2–5 years, 5–7 years, 7–11 years, and 11–adolescence. Generally, the topics were consistent across age levels, but in some case specific topics that were relevant to a particular age period (e.g., day care in the preschool years or sexual problems in adolescence) were included at only one age period.

Categories were generated from three sources. First, the present survey is a partial replication of information requested in a survey of South Florida pediatricians (Toister & Worley, 1976). Sections in the Toister and Worley survey requesting pediatricians to generate three areas of behavioral questions asked by parents for each of five age groups, questions concerning the extent to which behavioral problems are represented in the pediatricians' practices, and requests for suggestions concerning behavioral training for pediatricians were repeated in the current survey. Second, requests for developmental and behavioral bibliographies and curricula were made of 11 major medical schools in the United States. Topics addressed in these curricula and bibliographies were included in the present survey. Third, the senior author spent approximately 12 weeks accompanying two pediatricians on daily hospital rounds and office visits. Records were kept noting frequencies of episodes in which the pediatricians were required to have knowledge of developmental information.

For each topic in the survey, respondents were required to indicate on a 4-point scale (1 = high), the extent to which they would like to have learned more about a particular topic during training.

7.2. Results

The pediatricians in the sample nearly unanimously (98.7%) endorsed the importance of formal training in principles of behavior and child development for pediatricians in training. However, nearly 60% of the pediatricians had received no formal training in child development and behavior in medical school and 70% had received no such training—even in their pediatric residency. This is consistent with results of a questionnaire study of pediatricians, seventh-grade children, and their teachers and families that indicate that pediatricians rate themselves as less competent in dealing with psychosocial problems than in traditional medical areas. The perceptions of physician competence in these areas by seventh-grade children, their parents, and their teachers was positively related to the ratings of these physicians (Bergman & Fritz, 1980). The current importance of behavioral and developmental training for pediatric practice was indicated by the fact that the respondents reported an average of 21% of the daily office and hospital visits involved behavioral questions.

The survey revealed clear-cut patterns of pediatrician interests in child's age groups and in topics. In turn, the particular behavioral topics that pediatricians found of greatest interest interacted with age. The statistical significance of these patterns were evaluated by 5 × 10 analyses of variance with age period (5 levels) and topic (10 levels) as factors.

There was a significant linear trend for age periods as shown in Table 1, $F(4,268) = 3.72$, $p < .01$. Specifically, pediatricians indicated that they desired to learn more about developmental issues concerning children in the birth to 2-year age period and less across the later periods, with adolescence being of lowest interest. It should be noted, however, that even for the adolescent period, the mean response rating was 2.00 on a 4-point scale,

TABLE 1. Level of Pediatrician Interest as a Function of Age of Child

Age levels	Mean interest
Birth–2 yrs.	1.85
2–5 yrs.	1.90
5–7 yrs.	1.94
7–11 yrs.	1.95
11–adolescence	2.00

which indicated that the majority of pediatricians were still interested in learning more about developmental issues even for this age group. The results for topics of interest are presented in Table 2. Analysis of variance yielded a significant main effect, $F(7,469) = 2.68$, $p<.01$, for interest in topic. Developmental testing was rated by the respondents as of highest interest, followed by behavior problems and parent-child relationships, which were rated equally. Topics such as temperament, child competence, the vulnerable child syndrome, and child abuse were of relatively less interest.

However, as Table 3 indicates, these age and topic main effects were qualified by an age × topic interaction of borderline significance, $F(28,1876) = 1.35$, $p = .103$. Interest in some topics showed marked drops from the youngest to the oldest age periods, whereas interest in other topics showed stability across age. Specifically, interest in behavior problems showed a drop from the youngest age period (birth–2 years) to the third age period (5–7 years), but interest was stable from this later period through adolescence. Interest in the topics of parent-child relationships and child competence were relatively stable across all age periods. In contrast, interest in temperament,

TABLE 2. Level of Pediatrician Interest as a Function of Topic

Topic	Mean interest
Behavior problems	1.87
Developmental testing	1.80
Parent-child relationship	1.87
Child competence	1.92
Temperament	1.98
Vulnerable child syndrome	2.06
Nutrition	1.94
Child abuse	2.00

TABLE 3. Level of Pediatrician Interest as a Function of Age of Child and Topic

Age of child	Topic							
	Behavior problems	Developmental testing	Parent-child relationships	Child competence	Temperament	Vulnerable child syndrome	Nutrition	Child abuse
Birth–2 yrs.	1.77	1.74	1.82	1.92	1.83	1.86	1.91	1.93
2–5 yrs.	1.89	1.80	1.87	1.85	1.93	2.04	1.87	1.96
5–7 yrs.	1.95	1.73	1.85	1.95	2.04	2.10	1.93	2.00
7–11 yrs.	1.88	1.79	1.94	1.90	2.00	2.13	1.98	2.03
11–adolescence	1.88	1.94	1.90	1.96	2.08	2.15	2.01	2.08

the vulnerable child syndrome, nutrition, and child abuse showed relatively consistent and systematic decreases from the youngest age period through adolescence. Whereas pediatricians expressed greatest overall interest in the topic of developmental testing, there was a relative decrease in interest in this topic for the adolescent period, although this topic was still rated third highest by the respondents during the adolescent period.

7.2.1. Replication of Toister and Worley (1976)

In addition to this quantitative information, questions were included that tapped "the major areas of behavioral questions asked by parents about their children" at 5 age levels (birth–2 years; 2–5 years; 5–7 years; 7–11 years; and 11–adolescence). This question replicates closely the central behavioral question in Toister and Worley's South Florida survey. In their survey, Toister and Worley found that toilet training (all ages) and discipline (ages 2–11 years) were most often asked, with feeding and/or eating problems (birth–5 years), school fears (5–7 years), and school performance (7–11 years) being judged the next set of important problems.

In the present survey, there were no problems prevalent at all ages, but clear patterns were evident at each age. At the youngest age level (birth–2 years), three problems were most prominent: (a) eating disturbances, (b) management of infant crying, and (c) toilet training. Both eating and toilet training remained important problems at the 2– to 5–year period, and discipline also appeared at this time point. At the 5– to 7–year period, school and peer adjustment and hyperactivity appeared as problems, whereas discipline remained important. By 7–11 years, the same trio of problems remained: school adjustment, discipline, and hyperactivity. These problems changed little over the adolescent period, with school problems and discipline still major concerns followed by obesity and sexual problems.

The general pattern replicates the earlier findings of Toister and Worley with a few minor exceptions. Toilet training was a major concern in our sample mainly at the younger ages, whereas hyperactivity, obesity, and sexual problems were identified at older ages in the present study.

7.3. Implications

These results provide clear guidelines for curricula development for pediatrician training in child development. Indicated from our survey are an emphasis on the youngest age periods and a clear focus on developmental testing, parent-child relationships, and management of behavior problems.

Validation of the survey results through the use of physician records of actual parent inquiries as well as *in situ* observations of physician-patient interactions are necessary next steps. If these steps indicate that this interest survey technique is valid, it could serve as a highly practical and relatively inexpensive way of assessing the interests and needs of other professional groups in guiding curricula development for other specialized professional groups.

Next, a discussion of issues other than content associated with providing guidelines for the development of behavioral science programs in medical schools will be presented.

8. ISSUES IN TRAINING PEDIATRICIANS IN DEVELOPMENTAL PSYCHOLOGY

In view of our earlier argument that developmental psychological information should be made more available to pediatricians and in light of our survey results that indicate there is a perceived need for further knowledge and skills of this type, a final question needs to be raised, namely, how to provide most effectively developmental training to pediatricians.

The most direct approach is the introduction of developmental psychology into the medical school curriculum. An extensive literature has developed concerning alternative methods of training physicians. Briefly, three issues have been raised but have by no means been resolved.

First, which phases of physician training can best be modified to provide developmental and behavioral information? Some argue in favor of continuous presentation of this information throughout the curriculum (e.g., Pattishall, 1973; Stone, Gentry, Matarazzo, Carlton, Pattishall, & Wakely, 1977), whereas others favor one particular phase of physician training as especially relevant for the material; residency is most often suggested as the optimal phase (e.g., Lewis & Colletti, 1973; Rothenberg, 1968). At present, there has been no systematic evaluation of the effectiveness of these alternatives for maximizing the abilities of pediatricians in learning this type of material.

A second issue concerns the optimal curriculum format for presentation of developmental material, such as lecture course versus project-oriented course, seminar, or case-presentation approach (cf. Asken, 1979; Dacey & Wintrob, 1973; Lewis & Colletti, 1973; Parmelee, 1982; Sheldrake, 1973).

A third issue concerns the role of behavioral faculty in such training (cf. Chamberlin, 1973; Shonkoff et al., 1979; Stone et al., 1977; Wexler, 1976). None of these issues has been resolved at present, and in light of the increasing interest in introducing developmental/behavioral information into medical school training, further systematic evaluation of these issues would be worthwhile.

There are, however, other ways in which psychologically relevant material can be made available to pediatricians. The participation of psychologists in (a) programs of continuing education for physicians (e.g., Reiss & Hoffman, 1979, on the family; Kagan, 1980, on psychology's relevance for pediatrics) and (b) reviews of recent topics of interest to physicians (e.g., Taylor, 1979, on parent-infant relationships) are two ways that developmental psychological information is becoming increasingly accessible to pediatricians. The rise in specialty journals (e.g., *Journal of Pediatric Psychology* and *Journal of Developmental and Behavioral Pediatrics*) and in series devoted to the integration of pediatrics and behavioral sciences (e.g., *Theory and Research in Behavioral Pediatrics*) is encouraging. Moreover, the introduction of special interest areas devoted to this interface in organizations of both pediatricians (e.g., the American Academy of Pediatrics special interest area in behavioral pediatrics) and psychologists (e.g., the American Psychological Association Division of Health Medicine; the Society for Pediatric Psychology) are other signs of increased mutual interest. Finally, the recent emergence of a new professional organization, the Society for Developmental and Behavioral Pediatrics, a group organized to be interdisciplinary, attests to new opportunities for the bidirectional flow of information between these two groups. The bidirectional nature of this relationship should be emphasized; although the focus in this review has been on the perceived need of pediatrics for further developmental behavioral information, the complementary need of developmental psychologists for further medical, especially pediatric information merits increased attention. A complementary assessment of developmental psychologists need for pediatric information would be worthwhile and would serve to underline the bidirectional and interdependent nature of the emerging relationship between psychology and pediatrics.

8.1. A Final Comment

Although in the later sections of this chapter we have focused on the adequacy of psychosocial and developmental training for pediatricians, it is important to note that concerns have been expressed about the implementation of such training in standard pediatric practice. It is not clear that the relatively high interest and activity taking place in medical academic settings with regard to psychosocial and behavioral issues are being reflected in pediatric clinics and private offices. For example, in their editorial capacity at the *Journal of Developmental and Behavioral Pediatrics*, Gottlieb, Williams, and Zinkus (1983) suggest that in reality the commitment of the general pediatrics practitioner may not be adequate for pediatric provision of servces in the psychobiosocial area. It remains to be seen whether recent and current interest in developmental, behavioral, and psychosocial issues exists as a medical school training and research area or becomes more widely translated into everyday pediatric practice.

9. SUMMARY AND CONCLUSIONS

A historical review of the relationship between developmental psychology and pediatrics has been presented, including an exploration of the effects of the biomedical model, the evolving role of the pediatrician, and the role of psychology in pediatrics. Alternative strategies for strengthening the link between developmental psychology and pediatrics have been examined. Following a discussion of pediatricians' need for developmental information, the results of a national survey assessing pediatricians' interest in specific types of developmental information have been presented.

Effective pediatrician intervention in behavioral, psychosocial, and developmental areas will necessitate marked revisions in pediatric skills and attitudes, only possible with significant refocusing of pediatric education (Bergman & Fritz, 1980).

The results of the present survey and the accompanying discussion of other related issues concerning behavioral science curricula development for medical schools provide guidelines for curricula development for pediatrician training in child development. In terms of content, results from the present survey suggest emphasis on the youngest age periods and a clear focus on developmental testing, parent-child relationships, and management of behavior problems.

Other curricula considerations, such as a timing, format, and structure of behavioral science programs in medical schools, need further exploration and empirical investigation in order to best evaluate strategies for presentation of behavioral science material in the medical setting.

Richmond (1975) suggests that "mastery of child development in pediatric practice and training is an idea whose time has arrived" (p. 517). Although this has not always been true, the present review suggests that both disciplines are more accepting of what each has to offer. The next step is modification of educational and training programs to match the interdisciplinary objectives of collaborative research and application.

Acknowledgments

Our thanks to the staffs of Mercy Hospital, Urbana, Illinois, and of Burnham City Hospital, Champaign, Illinois, for their cooperation in the completion of this study. Thanks are also due to pediatricians Ronald Deering and Robert Boucek, for their advice and consultation on this project, and to the following individuals, for their helpful comments on earlier drafts of the chapter: Steven Asher, Orville G. Brim, Jr., Carol Eckerman, Robert Haggerty, Hiram Fitzgerald, Jerome Kagan, John Kennell, Claire Kopp, Arthur Parmelee, Stephen Porges, Donald Routh, Arnold Sameroff, Heidi Sigal, Michael Yogman, and Philip Zelazo.

We are grateful to Cathy Morris and the University of Illinois Survey Research Laboratory for help with data analysis and to Elaine Fleming for library research. Finally, thanks to Eileen Posluszny and Freda Weiner for their assistance in the preparation of the manuscript.

10. REFERENCES

Ainsworth, M. S. The development of attachment. In B. Caldwell & H. Ricciuti (Eds.), *Review of child development research* (Vol. 3). Chicago: University of Chicago Press, 1973.

American Academy of Pediatrics. Statement of objectives. *Operations manual.* New York: American Academy of Pediatrics, 1960. (p. 137)

American Psychological Association Task Force on Health Research. Contributions of psychology to health research: Patterns, problems, potentials. *American Psychologist,* 1976, *31,* 263–274.

Asher, S. R., & Gottman, J. M. (Eds.). *The development of children's friendships.* New York: Cambridge University Press, 1981.

Asken, M. J. Medical psychology: Toward definition, clarification, and organization. *Professional Psychology,* February 1979, pp. 66–73.

Belsky, J., & Steinberg, L. D. The effects of day care: A critical review. *Child Development,* 1978, *49,* 929–949.

Bemis, K. Current approaches to the etiology and treatment of anorexia nervosa. *Psychological Bulletin,* 1978, *85,* 593–618.

Benton, A. L., & Pearl, D. (Eds.). *Dyslexia: An appraisal of current knowledge.* New York: Oxford University Press, 1978.

Bergman, D. A., & Fritz, G. K. "The new morbidities." Physician competence and consumer utilization. *Journal of Developmental and Behavioral Pediatrics,* 1980, *1,* 70–73.

Blum, L. H. Some psychological and educational aspects of pediatric practice: A study of well-baby clinics. *Genetic Psychology Monographs,* 1950, *41,* 3–97.

Brazelton, T. B. *Neonatal behavioral assessment scale.* London: Heineman, 1973. (National Spastics Society Monograph)

Burnett, R. D., & Bell, L. S. Projecting pediatric practice patterns: Report of the survey of the Pediatric Manpower Committee. *Pediatrics,* 1978, *62,* 625–680.

Castelnuovo-Tedesco, P. How much psychiatry are medical students really learning? *Archives of General Psychiatry,* 1967, *16,* 668–675.

Chamberlin, R. W. New knowledge in early child development: Its importance for pediatricians. *American Journal of Diseases in Children,* 1973, *126,* 585–587.

Chamberlin, R. W. Management of preschool behavior problems. *Pediatric Clinics of North America,* 1974, *21,* 33–47.

Dacey, M. L., & Wintrob, R. M. Human behavior: The teaching of social and behavioral sciences in medical schools. *Social Science and Medicine,* 1973, *7,* 943–957.

Davis, J. K., Stone, R. K., Levine, G., & Stolzenberg, J. *Child psychology and pediatric medicine: A training program for primary care physicians.* Unpublished manuscript, New York Medical College, 1980.

Duff, R. S., Rowe, D. S., & Anderson, F. P. Patient care and student learning in a pediatric clinic. *Pediatrics,* 1973, *50,* 839–846.

Dworkin, P. H., Shonkoff, J. P., Leviton, A., & Levine, M. D. Training in developmental pediatrics: How practitioners perceive the gap. *American Journal of Diseases of Children,* 1979, *133,* 709–712.

Eisenberg, L. The relationship between psychiatry and pediatrics: A disputation view. *Pediatrics,* 1967, *39,* 645.

Engel, G. L. The need for a new medical model: A challenge for biomedicine. *Science,* 1977, *196,* 129–136.

Engel, G. L. The clinical application of the biopsychological model. *American Journal of Psychiatry,* 1980, *137,* 535–544.

Field, T., Sostek, A., Goldberg, S., & Shuman, H. H. (Eds.). *Infants born at risk.* New York: Spectrum, 1979.

Fitzgerald, H. E., Lester, B. M., & Yogman, M. W. (Eds.). Editors' introduction. *Theory and research in behavioral pediatrics* (Vol. 1). New York: Plenum Press, 1981.

Friedman, S. B. Foreword to symposium on behavioral pediatrics. *Pediatric Clinics of North America,* 1975, *22,* 515–516.

Gesell, A. Letter. *Pediatrics,* 1951, *8,* 734.

Gottlieb, M. I., Williams, J. E., & Zinkus, P. W. Exploring the roots of apathy. *Journal of Developmental and Behavioral Pediatrics,* 1983, *4,* 1–2.

Graham, F. K., Pennoyer, M. M., Caldwell, B. M., Greenman, M., & Hartman, A. G. Relationship between clinical status and behavioral test performance in a newborn group with histories suggesting anoxia. *Journal of Pediatrics,* 1957, *50,* 177–189.

Green, M. Coming of age in general pediatrics. Paper presented the meeting of the American Academy of Pediatrics, Aldridge Award Address, New York, October 1982.

Haggerty, R. J. The task force report. *Pediatrics,* 1979, *63,* 935–937.

Hansen, M. F., & Aradine, C. R. The changing face of primary pediatrics: Review and commentary. *Pediatric Clinics of North America,* 1974, *21,* 245–256.

Harper, P. A. The pediatrician and the public. *Pediatrics,* 1951, *7,* 430–432.

Haskins, R., Collier, A. M., Ramey, C., & Hirschbiel, P. O. The effect of mild illness on habituation in the first year of life. *Journal of Pediatric Psychology,* 1978, *3,* 150–155.

Haskins, R., Hirschbiel, P. O., Collier, A. M., Sanyal, M. A., & Finkelstein, N. W. Minor illness and social behavior of infants and caregivers. *Journal of Applied Developmental Psychology,* 1982, *2,* 117–128.

Helfer, R. E., & Kempe, C. H. *The battered child.* Chicago: University of Chicago Press, 1974.

Hinde, R. A., & Stevenson-Hinde, J. (Eds.). *Constraints on learning: Limitations and predispositions.* London: Academic Press, 1973.

Hornberger, R., Bowman, J., Greenblatt, H., et al. *Health supervision of young children in California.* Berkeley, Calif.: Bureau of Maternal and Child Health, State of California, Department of Public Health, 1960.

Kagan, J. The new marriage: Pediatrics and psychology. *American Journal of Diseases in Children,* 1965, *110,* 272–278.

Kagan, J. Overview: Perspectives on human infancy. In J. Osofsky (Ed.), *Handbook of infant development.* New York: Wiley, 1979.

Kagan, J. *The contributions of developmental psychology to pediatrics.* Paper presented at a conference on the Future of Academic Pediatrics, New Orleans, June 1980.

Kagan, J., Kearsley, R. B., & Zelazo, P. R. *Infancy: Its place in human development*. Cambridge: Harvard University Press, 1978.

Kearsley, R. B., & Sigel, I. E. (Eds.). *Infants at risk: Assessment of cognitive functioning*. Hillsdale, N. J.: Lawrence Erlbaum, 1979.

Kempe, C. H. The future of pediatric education: 1978 presidential address to the American Pediatric Society. *Pediatric Research*, 1978, *12*, 1149.

Kopp, C. B. Risk factors in development. In M. Haith & J. Campos (Eds.), *Infancy and the biology of development*, Vol. II. P. Mussen (Ed.) *Manual of child psychology*. New York: Wiley, 1983.

Lamb, M. E. (Ed.). *The role of the father in child development*. New York: Wiley, 1976.

Levine, S. Pediatric education at the crossroads: Presidential address, American Pediatric Society. *American Journal of Diseases of Children*, 1950, *106*, 12.

Lewis, M., & Colletti, R. B. Child psychiatry teaching in pediatric training: The use of a study group. *Pediatrics*, 1973, *52*, 743–744.

Lewis, M., & Rosenblum, L. A. (Ed.). *Friendship and peer relations*. New York: Wiley, 1975.

Littman, B., & Parmelee, A. H. Medical correlates of infant development. *Pediatrics*, 1978, *61*, 470–474.

Matarazzo, J. D., & Daniel, R. S. Psychologists in medical schools. *Neuropsychiatry*, 1957, *4*, 93–107.

McClelland, C. Q., Staples, W. I., Weisberg, I., & Bergen, M. E. The practitioner's role in behavioral pediatrics. *Journal of Pediatrics*, 1973, *82*, 325–331.

Melamed, B. G. Management of childhood disorders. In B. G. Melamed & L. J. Siegel (Eds.), *Behavioral medicine: Practical applications in health care*. New York: Springer, 1980.

Nathan, R. G., Lubin, B., Matarazzo, J. D., & Persely, G. W. Psychologists in schools of medicine: 1955, 1964, and 1977. *American Psychologist*, 1979, *34*, 622–627.

Page, H. E., & Passey, G. E. The role of psychology in medical education. *American Psychologist*, 1949, *4*, 405–409.

Parke, R. D. *Fathers*. Cambridge, Mass.: Cambridge University Press, 1981.

Parmelee, A. H., Jr. Teaching child development and behavioral pediatrics to pediatric trainees. *SRCD Newsletter*, Fall 1982.

Parmelee, A. H., Kopp, C. B., & Sigman, M. Selection of developmental assessment techniques for infants at risk. *Merrill-Palmer Quarterly*, 1976, *22*, 177–201.

Pattishall, E. G. Basic assumptions for the teaching of behavioral science in medical schools. *Social Sciences and Medicine*, 1973, *7*, 923–926.

Ramey, C. T., MacPhee, D., & Yeates, K. Preventing developmental retardation: A systems theory approach. In L. Bond & J. Joffe (Eds.) *Primary prevention of psychopathology*. Hanover, N.H.: University Press of New England, 1982.

Reiss, D., & Hoffman, H. (Eds.). *The American family: Dying or developing*. New York: Plenum Press, 1979.

Ricciuti, H. Developmental consequences of malnutrition in early childhood. In M. Lewis & L. Rosenblum (Eds.), *The uncommon child: Genesis of behavior* (Vol. 3). New York: Plenum Press, 1980.

Richmond, J. B. An idea whose time has arrived. *Pediatric Clinics of North America*, 1975, *22*, 517–523.

Richmond, J. B. Child development: A basic science for pediatrics. *Pediatrics*, 1967, *39*, 649–658.

Richmond, J. B. & Janis, J. A perspective on primary prevention in the earliest years. *Children Today*, May–June 1980, pp. 2–6.

Rogers, D. E., Blendon, R. J., & Hearn, R. P. Some observations on pediatrics: Its past, present and future. *Pediatrics*, 1981, *67*, 776–884.

Rothenberg, M. B. Child psychiatry-pediatrics liaison: A history and commentary. *Journal of Child Psychiatry*, 1968, *7*, 492–509.

Routh, D. K. Psychological training in medical school departments of pediatrics: A survey. *Professional Psychology*, 1970, *1*, 469–472.

Routh, D. K. Psychological training in medical school departments of pediatrics: A second look. *American Psychologist*, 1972, *27*, 587–589.

Routh, D. K. Research training in pediatric psychology. *Journal of Pediatric Psychology*, 1980, *5*, 287–293.

Sameroff, A. J. Can conditioned responses be established in the newborn infant: 1971? *Developmental Psychology*, 1971, *5*, 1–12.

Sameroff, A. J. Early influences on development: Fact or fancy? *Merrill-Palmer Quarterly*, 1975, *21*, 267–294.

Sameroff, A. J. Development and the dialectic: The need for a systems approach. In W. A. Collins (Ed.), *Minnesota Symposium on Child Psychology* (Vol. 15.). Hillsdale, N.J.: Lawrence Erlbaum, 1982.

Schaffer, H. R. *The growth of sociability.* Harmordsworth, England: Penguin Books, 1971.

Schaffer, H. R. *Mothering.* Cambridge: Harvard University Press, 1977.

Senn, M. J. E. Insights on the child development movement in the United States. *Monographs of the Society for Research in Child Development*, 1975, *40* (Serial No. 161).

Shea, V., & Fowler, M. G. Parental and pediatric trainee knowledge of development. *Journal of Developmental and Behavioral Pediatrics*, 1983, *4*, 21–25.

Sheldrake, P. Behavioural science in the medical curriculum. *Social Science and Medicine*, 1973, *7*, 967–973.

Shonkoff, J. P., Dworkin, P. H., Leviton, A., & Levine, M. D. Primary care approaches to developmental disabilities. *Pediatrics*, 1979, *64*, 506–514.

Siegel, L. J., & Richards, C. S. Behavioral intervention with somatic disorders in children. In D. Markolin II (Ed.), *Child behavior therapy.* New York: Gardner Press, 1978.

Sigman, M., & Parmelee, A. H. Longitudinal evaluation of the preterm infant. In T. M. Field (Ed.), *Infants born at risk.* New York: S. P. Medical and Scientific Books, 1979.

Sroufe, L. A., & Waters, E. Attachment as an organizational construct. *Child Development*, 1977, *48*, 1184–1199.

Stabler, B. Emerging models of psychologist-pediatrician liaison. *Journal of Pediatric Psychology*, 1979, *4*, 307–313.

Starfield, B., & Borkowf, S. Physician's recognition of complaints made by parents about their children's health. *Pediatrics*, 1969, *43*, 168–172.

Stolzenberg, J., & Levine, G. F. Integration of the developmental/behavioral science curriculum into the total residency program in primary care pediatrics. Paper presented at the meeting of the American Psychological Association, September 1979.

Stone, G. C., Gentry, W. D., Matarazzo, J. D., Carlton, P. L., Pattishall, E. G., & Wakely, J. H. Teaching psychology to medical students. *Teaching of Psychology*, 1977, *4*, 111–115.

Taylor, P. M. (Ed.). *Seminars in perinatology*, 1979, *3*.

Thomas, A., Chess, S., Birch, H., Hertzig, M., & Korn, S. *Behavioral individuality in early childhood.* New York: New York University Press, 1963.

Toister, R. P., & Worley, L. M. Behavioral aspects of pediatric practice: A survey of practitioners. *Journal of Medical Education*, 1976, *51*, 1019–1020.

Tuma, J. M. Pediatric psychologist . . . ? Do you mean clinical *child* psychologist? *Journal of Clinical Child Psychology*, 1975, *5*, 9–12.

Tuma, J. M., & Schwartz, S. Survey of consultation training at the internship level. *Journal of Clinical Child Psychology*, 1978, *7*, 49–54.

Wagner, N. Psychologists in medical education: A 9-year comparison. *Social Sciences and Medicine*, 1968, *2*, 81–86.

Werner, E. E., Simonian, K., Bierman, J. M., & French, F. E. Cumulative effect of perinatal complications and deprived environment on physical, intellectual and social development of preschool children. *Pediatrics*, 1967, *39*, 480–505.

Werry, J. S., & Sprague, R. L. Hyperactivity. In C. Costello (Ed.), *Symptoms of psychopathology*. New York: Wiley, 1970.

Wexler, M. The behavioral sciences in medical education: A view from psychology. *American Psychologist*, 1976, *31*, 275–283.

Whalen, C., & Henker, B. Psychostimulants and children. A review and analysis. *Psychological Bulletin*, 1976, *83*, 1113–1130.

Witmer, L. The common interests of child psychology and pediatrics. *Pediatrics*, 1896, *1*, 15–19.

Work, H. A. Exploratory role of psychiatry in pediatric education. *Pediatrics*, 1955, *16*, 408–411.

Wright, L. The pediatric psychologist: A role model. *American Psychologist*, 1967, *22*, 323–325.

Wright, L. Psychology as a health profession. *The Clinical Psychologist*, 1976, *29*, 16–19.

Yogman, M. Child development and pediatrics: An evolving relationship. *Infant Mental Health Journal*, 1980, *1*, 89–95.

Zeskind, P. S., & Ramey, C. T. Malnutrition: An experimental study of its consequences on infants in two caregiving environments. *Child Development*, 1978, *49*, 1155–1162.

Zeskind, P. S., & Ramey, C. T. Prevention of intellectual and interactional sequelae of fetal malnutrition; A longitudinal, transactional and synergistic approach. *Child Development*, 1981, *52*, 213–226.

2

A Socio-ecological Model of Adaptive Behavior and Functioning

NANCY A. CARLSON AND THOMAS Z. CASSEL

1. BEHAVIORAL PEDIATRICS: THE SOCIO-ECOLOGY OF HEALTH CARE AND SOCIO-ECOLOGICAL RESEARCH

1.1. Introduction

Pediatrics may be viewed as the medical discipline most essentially behaviorally and developmentally oriented. The pediatrician follows the human organism through a host of transitions from uterine to extrauterine existence through childhood and adolescence. Throughout the sequence of awesomely complex transitions, the pediatrician must monitor and strive to facilitate the organism's successive adaptations. From the highly buffered uterine environment, the organism confronts the differing adaptive demands of progressively less buffered and more complex social and ecological environments. Through participation in a widening set of environments, the organism develops and increases in behavioral complexity.

Nancy A. Carlson • Center for the Study of Youth Development, Stanford University, Stanford, California 94305. **Thomas Z. Cassel** • Plymouth Center for Human Development, Northville, Michigan 48167. Much of the theory and research cited herein was supported in part by grant G00760214 and contract 300-75-0255 to the Institute for Family and Child Study, Michigan State University, from the Bureau of Education for the Handicapped, U. S. Office of Health, Education and Welfare (now Department of Education, Office of Special Education), Washington, D. C.

The journey of the developing person through a specific series of socio-ecological environments is considered to constitute the individual's developmental trajectory. The developmental trajectory is literally a pathway through and within environments; a pathway along which the individual encounters a series of adaptive demands. Human development is considered to be tied to human socio-ecological environments (as will be discussed more fully in Section 3). The relatively unformed human neonate takes on form through interactive participation in human environments. Our human gait, speech, and conduct in general bear the marks of the environments in which they develop. Comprehending our human form requires an understanding of the environments through which we have passed and in which we presently function. Thus, the specific characteristics of our interactively negotiated pathway through environments represent the essence of human development (Cassel, 1974, 1976).

This chapter is primarily theoretical, outlining a holistic conceptual perspective for behavioral pediatrics; however, a few examples have been included to give the reader at least some idea of how the translation of theory into practice might be effected. Further, a few specific, but diverse, research questions have been sketched. More than anything else, this chapter offers an opportunity to approach pediatrics and the care of children and families from a different outlook.

The socio-ecological paradigm presented in this chapter has emerged through our attempts to grapple with the complexities of real-life environments; the worlds of developing children, their families, and the institutions to which they are joined. It has sought to benefit from a critical analysis of the historical trends in philosophy and the physical and life sciences. The terms and concepts employed were derived through a long process of negotiation, during which we strove to steer a course between arcane jargon and traditional terms with unacceptable connotations. The results of our endeavors are to be judged in terms of their fruitfulness in providing understanding and guiding research.

1.2. Family Systems and Behavioral Pediatrics

The system into which an individual is born or is early-on located becomes that individual's primary system. Through the establishment of linkages to the primary system, the individual receives support for survival and development. The primary system in the human may thus be the environment of the biological parents, a nonbiological family, or an institution. In whatever form, the primary system is the "launching pad" for an individual's developmental trajectory. The individual takes on his or her initial behavioral form and functional adaptations in relation to the interactive and ecological structure of the primary system. In addition, this system provides access links for transitions to other environments.

The additional systems to which the primary system is linked are adjunct systems. They are adjunctive in that they "add to" the support for survival and development of the primary system, and provide the additional environments necessary for continued development (Carlson, Lafkas, & Cassel, 1976). As Sullivan (1953) notes, "The change from newborn to infantile is less dramatic than its predecessor, but is a very great change indeed; and also entails a change of medium implying a change of functional activity in a complex of changing organization" (p. 33). An understanding of progressive changes in functional activity thus requires an explicit analysis of both primary and adjunct environments: the complex mediums of human existence.

From this holistic conceptual perspective, behavioral pediatrics can be characterized as a discipline concerned with understanding and facilitating the organism's behavioral transformations as it accomplishes adaptations to and transitions across an ever-widening set of socio-ecological environments. Behavioral pediatrics can simultaneously include a concern with the physiological processes that mediate functional activity and behavioral adaptation, and the characteristics of the environments in relation to which behavioral adaptation takes place. The challenge is to view the individual "patient" neither in isolation nor only from a clinical perspective; rather to view the individual as a functional human within and across a set of differing environments.

In addition, it is suggested that the socio-ecological perspective provides not only a blueprint for behavioral research on pediatric patients and their families, but also a blueprint for behavioral research on pediatricians, their training environments, and professional environments. We suggest that these two sets of concerns are necessarily interwoven. Hence, fruitful research in behavioral pediatrics must simultaneously address the behavior and the environments of children and their families *and* the behavior and the environments of physicians and other members of the health care delivery system.

1.3. Pediatricians and a Systems Approach

If one were briefly to sketch the activities of the clinical pediatrician, it would become apparent that he or she is called on to take significant responsibility for integrative management of children and families across a variety of environments over extended periods of time. Initially, the pediatrician is called on to help negotiate the transition of the mother and infant into a necessarily reorganized family environment. The relation of this negotiated entry to the behavior of all members of the primary system is a research issue with direct clinical significance. The work of Klaus and Kennell (1976) with delayed-return premature infants demonstrates the effect on the entire family system of obstructions to this initial process of reentry. On a less extreme level, it could be expected that many early clinical problems with feeding, sleeping, and respiratory stabilization are related to the manner in which the transition and adaptation to the family environment are negotiated. Later in the child's development, the pediatrician is frequently called on to address physiological and behavioral problems arising when the transition to school is not successfully negotiated. In these and other areas, basic research can furnish at least some of the information required for the pediatrician to provide anticipatory guidance and/or to work with members of the primary system to ameliorate potential difficulties.

In the case of illness, the pediatrician's direct responsibility for integrative management across contexts is also of interest. The pediatrician is called upon to guide and facilitate the child patient, with and without the family's support in accomplishing transitions from the home environment to an ambulatory setting, to diagnostic and hospital settings, and then to the home environment again. In that the child's functioning and "recovery" are considered to be directly related to the stress of transitions and the adaptation to environments, these latter factors require attention equal to the traditional concern with disease entities. Thus, a significant research concern for behavioral pediatrics is constituted by the analysis and facilitation of successful transitions and adaptations between and within environments.

A concern for successful transitions and adaptations can be noted in the following example. In the construction of a new outpatient medical clinic at Michigan State University (MSU), the concept of a drop-in child care center for well children was invoked. Since the clinic might well be the first medical environment that many young children would be exposed to, child development specialists worked with pediatricians and other medical personnel to prepare a supportive child care environment. In the process, several specific and basic research questions, such as the following, were asked:

- What is the transition like for the child who is "dropped off" by a parent or caregiver?
- What is the average age of children who attend? Range?
- What is the average length of stay in the unit? Range?
- Could medical personnel enter easily to examine a child either overtly or unobtrusively? How?
- How and where would "problems" be handled?
- Could dramatic play equipment of the "hospital-type" (i.e. stethoscopes, syringes, and doctor's bag) decrease presumed anxiety?
- What types of educational play equipment are necessary?
- How much individual attention is needed by the child?

• How can a drop-in center be staffed efficiently and effectively?

These and other research questions regarding the specific child care/ medical environment and transitions have been asked during the past five years. Now, after more than 20,000 visits by children, a picture has emerged. Interestingly, it does appear that if the transitional environment is carefully orchestrated—at entry and at departure—there are very, very few (less than 1%) problems for children or adults. The majority of the carefully collected observational data has yet to be published; still, the concern of many individuals in creating and analyzing an environment for children in a medical complex is encouraging.

At the same time, the socio-ecological perspective delineates a research domain focusing on the behavior and environments of the pediatrician, since the pediatrician's training and functioning directly impinge on the therapeutic goals for his or her patients. Just over ten years ago, Brent and Morse (1969) succinctly pointed to a major problem in the adaptation of pediatricians. If their remarks are read with an openness to the notions of transitions and adaptation to multiple environments in mind, then the relation of socio-ecological research and training will become more apparent.

> No wonder pediatricians have insurmountable "re-entry" problems when leaving their residencies to practice pediatrics. They have not been trained for their jobs! They must start over, retrain, reorient, innovate. Only some survive the re-entry shock; the others initiate the migration from clinical pediatrics. (p. 788)

The socio-ecological perspective suggests that training in general is not essentially constituted by the acquisition of isolated skills or pieces of behavior. Rather, education and training are considered to represent a contextually situated developmental process; one in which an individual's conduct gradually comes into adaptive coherence with the structure and demands of specific socio-ecological environments. It differs from other situated developmental processes only in its requirement to generate a specific output product. For all developmental processes, adaptive behavior is the outcome of an achieved fit between the individual and the set of social and ecological structures defining the enveloping environment.

In that behavior is always environment specific, the degree to which the training environments match or simulate the target environments of professional functioning will predict the degree to which an individual will successfully negotiate the transition to adaptive functioning within them. For pediatricians, the hospital environments that constitute the major loci of training only match in a partial and truncated fashion the environments of clinical practice. It thus becomes clear why successful adaptation to the hospital environment will neither predict nor necessarily facilitate successful adaptation to other practice environments.

What is most incomplete, given only hospital-specific behavior on the part of the pediatrician, is an awareness of and sensitivity to the developmental trajectory that has brought the child to the hospital environment. Correlated with this is a relative unconcern with the organization of the child's transitions within the hospital environment, as well as the transition to and from the hospital. However, a concern with these factors is critical to the behavioral pediatrics of clinical practice. Again, a Michigan State example may help to clarify this point.

Several medical school administrators concerned with the functioning and adaptation of their graduates in pediatrics (as well as those in other branches of medicine) sought to introduce curriculum changes. In conjunction with child development specialists, they provided information about and some experience with "normal" children in early childhood centers. By introducing the concept of a developmental continuum and experiential practice with developing children, they sought to decrease the discrepancy between medical school training and the requirements of clinical practice in pediatrics. The administrators were hopeful that the behavioral interactions thus learned and practiced would generalize to sick and hospitalized children as well. Unfortunately, when there was space in the medical curriculum, the importance of the experience (perceived by third-year medical students) was low. When relevance was at its peak (i.e., pediatric residency) other curricular and practice demands were such that the guided experience with children was not all that it might have been. Development of children takes place in a time-space field that tends to exclude those in medical schools. Still, for some (particularly three or four residents), the experience and information were invaluable.

A pressing research requirement is the socio-ecological mapping of practice and training environments. The mapping process entails a precise specification of the ecological and social structures of each environment. In addition, the access links between environments and the manner in which transitions are accomplished require specification. Through this process, training and practice environments can be brought into greater coherence, and the enhanced adaptation of pediatricians may be fostered (Cassel, 1978).

In addition, through the results of the mapping process, the delivery of health care can be optimized. The mapping process will lead to an understanding of the relation between the structure of health care environments and health care outcomes. For example, under what conditions is it better to design (redesign) an architectually open waiting-room environment? At what ages are children aware of and/or responsive to, specific health care garb—white coats, and so forth? Changes in health care outcomes may be brought about through the research-based modification of health care environments.

Environmental modifications will give rise to changes in individual and in interpersonal behavior. And these changes are of interest, for the relation between health care behavior and the efficacy of the health care process is important to behavioral researchers. For example, in the drop-in child care center at MSU, it was discovered that children had much more difficulty with the departure transition: They did not want to leave the unit and/or go with their caretakers. Changes in behavior of health care and child care personnel were introduced, along with several environmental modifications. Leaving then became almost as much fun as coming to the unit, thus relieving some traumatic moments for "parents" concerned about their child's welfare. The structure of health care environments, health care behavior, and health care outcomes represent interdependent factors of the overall health care process. Thus, a broadened research paradigm is made available and medical concerns are seen to be linked to ecological and behavioral concerns.

The concerns of health care training and delivery converge in the direct clinical encounter between physician and patient. The direct face-to-face interpersonal encounter represents the "final common pathway" of the health care delivery system. Yet the clinical encounter may be one of the least investigated components of the overall health care process. Until very recently, our emphasis upon the technology of health care has obscured from view and placed in a questionable light the therapeutic significance of the inter-personal relationship between patient and physician.

Wolpe (1973) notes that the one common ingredient which may account for similar success rates across a diversity of treatment systems for neurosis is the relationship between therapist and patient. For Wolpe, this relationship functions specifically in regard to stress adaptation and inappropriate anx-iety-response habits. Recent research (Aschoff, Fatranska, Giedke, Doerr, Stamm, & Wisser, 1971; Cassel & Sander, 1975) suggests that biorhythmic entrainment to social events serves as the physiological basis of the attach-ment relationship. The effects of social entrainment on physiological func-tioning may allow an eventual understanding of both placebo effects and the efficacy of traditional "medicine men." In this light, the working alliance and relationship between a physician and a patient demands to be investigated.

A working alliance directed at the maintenance or reachievement of a state of healthy functioning requires the participation of both members of the interaction in a highly ordered series of interactional events. The effec-tiveness of the medical encounter and its therapeutic sequelae are directly constrained or assisted by the degree to which an interactive alliance and information feedback loop can be established. Procedures similar to the inter-active research methods developed to analyze the mother-child relationship (e.g. Brazelton, Koslowski, & Main, 1974; Condon and Sander, 1974; Schaf-fer, 1977; Stern, 1974) may now be utilized to investigate the genesis and maintenance of the physician-patient relationship. For example, a comparative longitudinal videotape study of the factors leading to the development of differentially effective therapeutic relationships between pediatricians and their parent/families would constitute basic research into the processes of human interaction and applied programmatic research (Cassel, 1978).

One of the distinguishing characteristics of the socio-ecological perspective that recommends its value and usefulness is its theoretical insistence on the linkage between basic research and real-life environments. From this perspective, research is to be conducted either in naturally occurring environments or in controlled laboratory settings that explicitly model significant socio-ecological characteristics of target environments. By means of this approach, the ecological validity of research may be directly ascertained; and the direct translation of basic research to applied concerns may be immediately achieved.

In the following section a case history will be presented to portray (more concretely) the complex issues involved in the negotiation of an adaptive fit. Concepts raised in the present section will be employed to shed light on the processes through which a handicapped adopted child may achieve an adaptive fit within a family and community context.

2. NORMAN: AN ECOLOGICAL CASE STUDY OF A YOUNG CHILD AND HIS DEVELOPMENT

Norman, at 18 months, was very unlike his cohort of the same chronological age. His world was limited to the crib; his view of the world is and has been severely constrained. Norman's primary environment was impoverished and his functional activity reflected this impoverishment. His modes of environmental participation reflected the characteristics of the limited environments. Small wonder that with only ceiling, cracks, and cobwebs to stare at, he was thought to be "unable to relate."

When this story begins, Norman is in an institution for the profoundly mentally retarded. Although there had been no definitive assessment—few people went near Norman—most of his caretakers had assumed (for his short life) that he was a "vegetable." His primary system is a dead-end one, which is supportive only of survival and not of development. It has no access links to other environments, and no adjunct environments can be specified.

Norman was born of two institutionalized individuals who appeared to care deeply for each other. They had asked for permission to marry and had been refused. The officials in charge of the institution in which the biological parents resided felt they were "too retarded" to have children. Undeterred, they proceeded to conceive and have a child of their very own.

Although Norman appeared to be loved very much at birth, and appeared to be a member of a developing primary system, the institutional staff felt he would not be cared for appropriately. Thus began a cycle that was broken only by extreme courage and determination on the part of an unusual couple.

Norman at one month was taken from his biological parents and was placed in the first of a series of foster homes. The transition to this new environment was negotiated poorly. He was unable to achieve a fit with this family, and within a few weeks he was moved to another foster home. In all, he was moved into and out of fourteen foster homes in fourteen months. Had he even begun to establish connections with his environments, it would have been difficult if not impossible for these to have developed beyond a rudimentary state.

The last foster home was, perhaps, the worst. Someone (it is difficult to ascertain who was responsible) said he was having "seizures" and needed medication. In the absence of concern for the effect on his development, medication was prescribed, and Norman's presenting state was radically altered. He had begun his travels into a disoriented world; he would stay there for three months, unaware of the next transition—into an institution from which it was felt he would never emerge.

This child who was cut off from the world, who had repeatedly had potential connections to society ruptured, whose life pattern was one of displacement and isolation, was considered "unadoptable." Nevertheless, there were two special individuals who wanted Norman. Having traveled hundreds of miles from their home to an out-of-the-way institution, they believed Norman was meant for them. They wanted Norman to enter their childless family system. They had reserved and prepared a place for him.

But obstacle after obstacle confronted them; every day more discouraging news was presented by institutional personnel. Still they persevered. Finally, long after others would have given up in despair, they were able to begin the long journey home with Norman. He was to take his place in a warm, supportive context. They carefully supported the transition to his new primary environment and sensitively restructured it in light of his slowly emerging characteristics. The rhythms of their lives shifted into coherence with Norman's rhythms, and the ecology of their home was modified by his presence.

For the new family, however, the problems were not ended. They gave to Norman a loving environment filled with attention and mutually beneficial exchange processes. But Norman needed more than the connections and structure of a primary environment; he desperately needed professional services and necessary adjunct environments. The family then began more journeys, more searching for supportive adjunct systems—from town to town and center to center—in search of someone, some professional who would help them.

When Norman entered his new primary environment he was 18 months old, but was functionally a 3-month old child. He was unable to sit, swallow, chew, babble, or achieve eye contact. The previous absence of interactive participation in human environments was now reflected in the absence of behavioral modes of interaction and self-regulation. He lay on his back and stared at the ceiling: his only known and familiar context for the past 18 months. The new parents worked with Norman, establishing and maintaining supportive interactions, and gradually introducing increasing complexity. And yet they knew that, without help, successful adaptation and development were unlikely.

Both adoptive parents were employed; in addition, they lived in a small town, over 40 miles from the nearest city with sufficient resources for their needs. They were without any linkages to support systems and gradually began to feel totally disconnected from the rest of the world. There was no one within the health care agency network who would offer any alternative except reinstitutionalization. Not only Norman's but their own adaptation was at stake.

At the limit of their emotional strength, they finally found a pediatrician who listened and understood. "What you need," she said, "is an environment where there are 'normal' children around. Norman needs stimulation and appropriate peer modeling." But the family had already contacted over 40 child care centers and family day care homes in the area without any luck. No one would accept Norman. "Why don't you try the day care center where my son works?" the pediatrician asked. "They have made a commitment to children and families with special needs. Perhaps they would accept Norman on a trial basis, and I could provide assistance through my son."

The center accepted Norman and, since it was located on the campus of a major university, many resources were readily available. The center director and the infant unit coordinator saw the adaptive training of Norman and the supportive help to the family as an exciting challenge. The mother worked at the university (but had not known about the center) and was able to bring Norman to and from the center. Thus, she was able to guide and help Norman negotiate the transition to a new environment.

So began an enormous effort on the part of a large number of professionals, paraprofessionals, and students to provide for Norman the type of environment in which he would have the maximum opportunity to succeed (see Section 4). Norman blossomed. He began to chew, sit, stand, babble, imitate, walk, and say one or two words. In new and supportive developmental contexts, he began his long delayed development. To say that the family, staff at the center, and the pediatrician were delighted is to understate the case. Although Norman had not yet caught up with his peers, he was progressing by leaps and bounds. Successful development, as opposed to survival, was clearly a possibility.

.   Inspired by Norman's success, the family decided to adopt another handicapped child. After first checking that the same resources would still be available, they began their search for a second child. They soon found another young child with special needs and moved quickly to bring him into their supportive environment.

Jason was unlike Norman. He was able to eat, walk, and travel from place to place on his own. But he did not want anyone (adults or children) near him. Jason's functional adaptation to his previous environments was characterized by distancing behaviors and self-stimulation. The family's attempts to reach out to Jason were met with active avoidance—the more they approached, the more he withdrew. The family and the center staff were confused. Why was this happening? Why was Jason withdrawing further and further from their reach? Seemingly, they had the best resources available, but Jason did not fit. Ultimately, Jason was not adopted by this family who had so longed for another child. He returned to the adoption agency and was placed in another foster home, eventually to achieve a more successful adaptation.

Why had this happened? It is only now that the family and all the other individuals involved can begin to understand. The achievement of an adaptive fit within and between family, child, and environment is based on a long process of negotiation. There are some real constraints or barriers to this achievement. No system, no matter how willing, is infinitely flexible. Every system has its own constraints. The mutual modification for and with Norman could not be achieved with Jason. It is possible that the great desire to help and to be linked to Jason was itself a hindrance. Possibly in light of his previous developmental history, Jason required a larger measure and longer period of benign noninterference. With our present knowledge, it is difficult to be certain.

Norman has successfully achieved an adaptive fit. He has made the transition from a dead-end system to one in which he is vitally and essentially interactive. Norman will continue his developmental trajectory, and so will Jason; but Jason will need a different set of transactions and environments. As our knowledge of developmental processes increases, we will be able to more assuredly match and support children and environments. It is to be hoped that this pediatrician (and all pediatricians) will continue to be a resource in the lengthy and rather difficult process by which adaptations are continually negotiated.

The concepts represented in the story of Jason and Norman go beyond the notion of individual differences and thus pose intriguing research questions for behavioral pediatrics and other disciplines. But before discussing specific research questions, we shall examine briefly the theoretical basis of the socio-ecological perspective. The theory may serve as a guide to (but not necessarily a predictor of) understanding specific behavioral outcomes, as . we look at individuals within and across their varying contexts.

3. THE CONCEPTUAL BASIS OF A SOCIO-ECOLOGICAL THEORY

3.1. Converging Models in Philosophy, Physics, and Biology

We live in a most exciting time for theory and research in the life sciences. Traditional ways of thinking about phenomena in all the sciences are being shattered; novel models are coming to the fore. At the same time, new technologies are becoming available that allow for the precise investigation and recording of new phenomena.

The twentieth century was ushered in by the articulation of new and converging perspectives on physical and on living systems. The physical and life sciences had become hampered by an adherence to mechanistic and static paradigms. For both areas, more traditional notions are being replaced by paradigms that emphasize systems of interconnected processes situated in space-time fields.

An example of such a shift in the life sciences was Darwin's replacement of the static and ideal type view of the species with a contemporary perspective, viewing a species as a population spread out in space and time. It is a population in which all the members work to adapt to their surround; one in which all members, although unified by species characteristic morphologies, display unique individual differences across all measurable variables. Thus, there can be no single ideal species type, only diverse individual members of a species population. At the same time, Darwin emphasized that each population is situated in relation to a specific ecological context. Each individual member of the population adapts to its environment in a slightly different way. The members actively confront their environment and work to achieve an adaptive fit. The differential adaptation of members of a population and their offspring to an ecological context becomes the motor of evolution.

Whereas these notions might seem almost commonplace now, it is important to recognize how radical this change is; and that this new perspective has not as yet effectively penetrated the mainstream of the behavioral sciences. An emphasis on individual differences has typically not been found within the behavioristic nor cognitive perspectives, and hence the clinical realities of working with individuals may be ignored both in theory and in practice. The actual morphology of the organism may deserve significant attention. But only recently, under the influence of ethologists (Hinde & Stevenson-Hinde, 1973) and comparative psychologists (Kuo, 1967; Tobach, Aronson, & Shaw, 1971), have researchers concerned with human behavior begun to focus attention on morphology and the significance of morphological constraints.

The connection of the organism to its world is a notion central to the new currents not only in biology but also in physics and in philosophy. Specifically, behavior is most adequately understood as the patterned activity that results from the organism's linked participation with the contextual situation. The active, integrative form of the organism is linked with the form of the environment giving rise to the output of patterned activity. From this perspective, the explanation of behavior requires equal attention to the form of the organism, the form of the environment, and the exchange processes between the two. Since both the form of the organism and the form of the environment are continuously changing and unfolding, the output (behavior) will also be dynamic and changing.

The perspective of linked systems is central to recent advances in molecular biology. What is required is attention to the stereochemical form of macromolecules and attention to the coupling of each form to its medium. For example, it is both incorrect and misleading to state that DNA alone contains directive information. DNA molecules outside of an appropriate aqueous medium exhibit a denatured secondary form. When placed in an aqueous medium, their hydrophobic and hydrophilic couplings with the medium further twist the macromolecule into its active tertiary form. The medium and the coupling to it supply essential configurational information.

Similarly for behavior—it is not stored in the organism solely as behavioral routines or as cognitive structures. Behavior results from the organism's active linkage with its environment. It is in this sense that a precise map of the structure of the environment—the size, shape, and arrangement of rooms, and chairs, as well as the analysis of the social structures and interactions— is of equal importance to an adequate explanation of functional behavior, as is a precise account of the structure of the organism.

The same emphasis on the coupling of the organism to its surround is characteristic of contemporary physics. As the physicist and philosopher of science David Bohm notes, in order to understand a score of implications of Einstein's special theory it is necessary to realize

> that all of these considerations arise out of the need to take into account the important fact that *the observer is part of the universe*. He does not stand outside of space and time, and the laws of physics, but rather he has at each moment a

definite place in the total process of the universe, and must be related to this
process by the same laws that he is trying to study. (1965, p. 177, emphasis in
original)

And the universe to which the observer is necessarily connected is not
a container universe of isolated things or events. Rather, it is a space-time
field of interdependent processes. From this vantage point, the shared perspective of physics and of biology may be recognized. The revolutionizing
force of the theory of evolution is to be found in the interconnected web that
it spins between all forms of life and their surrounds—their necessary mediums of existence. It is a network of living systems spread across space and
back into time: a space-time field theoretic account.

The emphasis on the structure in time of events further weds contemporary physics and biology. For Newton, space and time were independent
and absolute metrics that characterized the container in which the elementary
particles of matter moved. For Einstein, space-time was conceived as a rela-
tivistic manifold; the space-time metrics were dependent on and constituted
by the event-processes themselves.

For biology, the overthrow of an absolute and independent metric of
time has been accomplished by recent work in biorhythms research. Temporal structure is now seen to be as constitutive a characteristic of a biological
system as is its structure in space. The biologist John Bonner begins his
treatise *On Development* by discussing the ubiquitous presence of developmental cycles across all levels of organization of living systems. Informational
macromolecules display life-cycle structures in time, as do organelles, cells,
and the organism with its life cycle (1974, pp. 11–15). The various levels of
cycles sustain complex phase relations with each other. Similarly, the behaviors of organisms are observed to display multirhythmic structures in time.
Daily circadian rhythms of behavior are well known, but in addition there
are more rapid infradian rhythms and more slow ultradian rhythms of behavior. In this light, a living system may be more adequately thought of as a
structured performance in time. It naturally follows that just as physicists
seek to map the space-time behavior of elementary particles, so we must seek
to map the space-time behavior of organisms—specifically humans.

The presence of a complex manifold of rhythms holds for physiological
as well as for behavioral processes. Hence, the former may also be thought
of as patterned temporal performances. It is a mistake to fundamentally
separate behavior from physiology; when this is done an unfortunate dualism
arises. With separation, either behavior is reduced to physiology (and with
this reduction the morphology of the organism [the body] and its environmental linkages somehow disappear) or behavior is cut free and let loose to
float in a cognitive or behaviorist space. It is thus necessary to recognize

that physiological processes *are* behavioral processes. They are output behavioral processes resulting from the linkages of suborganismic systems with their necessary mediums. In exactly the same way is "behavior" the output of the linkage of the whole organism to its environment. No level of linked functioning can be reduced to any other level. At the same time, no level can be understood apart from a consideration of cross-level feedback connections, and the constraints imposed on the functioning of a given level of organization by both higher and lower levels of organization.

John Dewey, possibly the most neglected of twentieth-century philosophers, clearly articulates both the indissoluble linkage of organisms to their environments and the temporal structure of this adaptive process.

> Whatever else organic life is or is not, it is a process of activity that involves an environment. It is a transaction extending beyond the special limits of the organism. An organism does not live *in* an environment; it lives by means of an environment. . . . The processes of living are enacted by the environment as truly as by the organism; for they *are* an integration. . . . It follows that with every differentiation of structure the environment expands. For a new organ provides a new way of interacting in which things in the world that were previously indifferent enter into life-functions. . . . As long as life continues, its processes are such as continuously to maintain and restore the enduring relationship which is characteristic of the life-activities of a given organism. . . . Indeed, living may be regarded as a continual rhythm of disequilibrations and recoveries of equilibrium. (1938, pp. 25, 26, and 27, emphasis in original)

The organism's environmental participation simultaneously expands its life-space and its developmental complexity. As the complexity of its environment increases, so the complexity of its organization increases. As we transform our world, we in turn are transformed.

3.2. A Socio-ecological Model of Adaptive Behavior

The present perspective on adaptive behavior is being developed in light of the foregoing conceptual models. It is being developed as a response and contribution to the contemporary search for paradigms in the behavioral sciences capable of embracing the novel currents that seek to alter traditional Western perspectives on the world and our place in it.

In contrast to a traditional view of ourselves as separated and disconnected, it offers a view of man as inextricably situated and connected. From this perspective, an observer and his or her observational instruments are connected to the products of active observation. The world comes to be known in the process of our operating on it.

In the search for a heuristically valuable paradigm, several currents have been woven into an empirical theory of the development of adaptive behavior. The view of a developing organism increasing in informational complexity has been joined to the socio-ecological perspective of recent primate studies to give rise to contextual developmental theory (Carlson, 1976; Cassel, 1976).

If development is defined as an increase in informational complexity, then the central question for any developmental theory is from whence comes the information? Nativist theories posit that the information is fully present in the genome. In this perspective—as in others—information is seen as arising from the linkage between the structural complexity of the organism and that of its environment. As is the case for the DNA macromolecule that takes on form through its environmental linkage, likewise the organism is viewed as being progressively informationally enriched through its linkages to its environments. Living systems may thus be viewed as order maintaining and order increasing machines. That part of their history in which their order increases is their developmental epoch (Cassel, 1972).

Organisms are linked to their environments by means of energy driven exchange processes. An increasingly complex variety of exchange processes link us to our environments: from oxygen-carbon dioxide exchange, to the exchange of daughters for bride payment, to the exchange of energy, and to the exchange of information with our environments in the processes of empirical investigation. In all instances of exchange processes, we are modified as we simultaneously modify our environments. Through processes of mutual modification, an organism comes to achieve a temporary fit with its surround. This bidirectional process of mutual modification is the basis of behavioral adaptation.

Processes of mutual modification are at the heart of all biological processes of adaptation. In general, the mutual fit between organisms and environments is considered to be the evolutionary result of the reciprocal structuring influences of the one on the other. "The miraculous fittingness of the various environments, and the interdependencies of all living systems is to be understood from this perspective. We are made for each other because we are made by each other" (Cassel, 1972).

In seeking to understand the emergence of developmental organization through social interaction, Louis Sander (1962, 1964, 1969, 1975) has described a paradigm for the investigation of adaptive fit. Sander was interested in following not only the physiological and behavioral developmental processes of the infant, but also the dyadic interactions of mother and infant, which appeared to lead to mutual modification ("interactive regulation").

Using a longitudinal study of 22 mother-infant dyads followed intensively over the first 18 months, Sander recognized both the uniqueness of each dyad as well as an interactive and temporal commonality. The quality of regulation of a dyad displayed a cyclic pattern: periods of stress would be followed by periods of harmonious integration, and particular regulatory and interactive concerns for one dyad tended to reappear on a similar timetable for other dyads. As Sander (1969) notes:

> This gave the impression that we were watching a sequence of adaptations, common to all the different mother infant pairs, although acted out somewhat differently by each. Each advancing level of activity which the child became capable of manifesting demanded a new adjustment in the mother-child relationship. . . . As one follows the mother infant pair from the beginning, one can represent the degree of harmonious coordination which will be reached in any one of the epochs as a particular open-ended question . . . an issue, which is in the process of being settled for each individual pair. (p. 191)

According to Sander, the process of resolving a developmental issue was a matter of interactive negotiation across all levels of organizations. Through a process of mutual modification, both members of the dyad negotiated a regulatory fit. For both members of the dyad this involved modification of behavior patterns and their biological rhythms. The negotiation process was at work whether one looked at the achievement of day-night differentiation during the first few months, or at the organization of the child's self-assertive navigations through his environment during the end of the 18-month period.

During the course of the first 18 months, Sander delineated the presence of five major adaptational issues. The manner in which the dyads negotiated each consecutive issue gave rise to the sequence of adaptations characteristic of that dyad. Following Sander, one can suggest that an adaptive fit is the temporary outcome of complex, ongoing processes of negotiation: a temporary equilibrium that must be periodically reachieved in the face of new, emerging issues.

However, while Sander has emphasized the interactive connections of the dyad and its structure in time, the processes of interactive negotiation must be seen as being profoundly situated processes. An understanding of behavior and interaction requires that it be viewed as necessarily situated—situated in, and linked to both social and ecological contexts. This view gives rise to an emphasis on the total context of conduct (Carlson, 1976).

The notion of total context of conduct derives from the work of a group of evolutionary biologists and ethnologists (Clutton-Brock & Harvey, 1977; Crook, 1965, 1970; Cullen, 1957; Eisenberg, Muckenhirn, & Rudran, 1972: Huxley, 1923) and underlies the contemporary socio-ecological perspective. Recognizing the inadequacy of an analysis restricted to the behavior of individual organisms, these researchers began studying the relation of individual conduct to the social and ecological context within which the behavior was situated. From this work, it has become clear that for group-living species, the conduct of an individual member is variously determined by, and adapted to, the conduct of the social group. In addition, the structure of the social group is coadapted to the ecological context inhabited. This research has demonstrated that closely related species living in divergent environments display strikingly different forms of social organization, whereas quite unrelated species living in similar environments display similar social systems.

In order to complete this socio-ecological perspective, a life-cycle component is required. Across the life-cycle, the organism sequentially enters and functions in relation to a widening series of progressively more complex socio-ecological environments. As an individual enters a new and different socio-ecological environment, new adaptive demands are posed and different forms of interaction develop. Different sets of developmental issues requiring interactive negotiation emerge with entrance to changing environments. Each socio-ecological environment is literally a developmental context. With the achievement of a negotiated linkage to a different environment, developmental complexity increases. Thus, access to new environments represents access to the contexts of continued development. Precluded access necessarily compromises development, although facilitated access enhances the achievement of adaptive functioning.

In the earlier example of Norman, the institutional setting severely limited access and thus opportunities for development, whereas a supportive home environment and access to other children offered almost unlimited opportunities for continuing development. Access, therefore, becomes a critical issue and one on which professionals can concentrate.

As we follow the human organism in its emergence from the buffered environment of the womb, we can map the developmental journey across linked socio-ecological environments. For the nonambulatory infant, access to new environments is mediated by the support function of the primary caregiver. As the human infant becomes ambulatory, his entrance into and navigation through environments is mediated and structured by older siblings and adults present in the home. As the child's world broadens, he navigates beyond the primary home environment and enters linked adjunct environments (Carlson *et al.*, 1976). New interactive partners, peers, nonfamilial adults, and new situations come rapidly one on the other.

Given this perspective, five basic aspects of the socio-ecological model of contextual development may be graphically represented. In Figure 1, a schematic map of the model components is presented. It is important to emphasize that the model may be seen as a map, for the perspective requires that we map the activities of organisms within and between environments. Theoretically, the methodological goal is the construction of a space-time map: the precise second-by-second and millimeter-by-millimeter

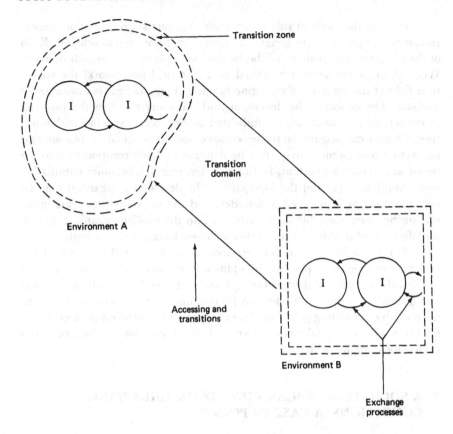

FIGURE 1. Socio-ecological model of contextual development.

documentation of the actually occurring activities of organisms within varying environments.

The model/map specifically directs us to the examination of (a) the characteristic of the organism across all levels of functioning—an expanded notion of organismic state; (b) the social and ecological state of an environment; (c) the exchange processes linking organism and environment; (d) the accessing relations between environments and cross-environment transitions; and (e) the temporal structure of the negotiation of an adaptive fit.

From this perspective, we define adaptive functioning as issuing from the achievement and maintenance of linkages between organism and environment. Their achievement and maintenance are seen as based on processes of mutual modification. Such processes are pictured as continuously accomplished, negotiated achievements produced by the organism-environment system.

It is on the basis of this perspective that our suggestions for theory, research, and programs in behavioral pediatrics have been developed. Each of the foregoing components of the model/map specifies a research domain. When these components are viewed in a historical framework, the space-time field of the complex of changing functional adaptations is available for analysis. The notion of the developmental trajectory provides the unifying construct for this analysis. As indicated in Section 1, specific and novel research projects focusing on the processes of socio-ecological adaptation and contextual development may then be delineated. Each component may be recast as a research question. In this manner, research becomes cumulative and complimentary; and the life-space of the developing organism may be ever more precisely mapped and understood. At the same time, this information becomes available for translation into the applied organization and modification of health care and other socio-ecological environments.

The pediatrician finds himself necessarily implicated in much of the trajectory of developing persons. In addition, the pediatrician is often involved in critical environmental transitions. To shoulder both a significant clinical and a research responsibility for this process may be seen as both a demanding and an exhilarating prospect. An example of a broad scope project, based on the socio-ecological perspective and incorporating varying research dimensions, might be helpful.

4. A SOCIO-ECOLOGICAL VIEW OF HANDICAPPING CONDITIONS: A CASE IN POINT

In the health care network, there are insufficient programmatic models based on a socio-ecological perspective. This is not surprising given the demands and complexity of such a model; but for those who might wish to implement and/or study this type of approach, the opportunity to examine successful models is generally not available. To help remedy this lack and illustrate how a seemingly abstract theory may be translated into a more definitive program, an example has been included. The program in which this model is imbedded is called PATHWAYS: A Human Support System Model for Integrated Handicapped Children and Their Families, usually referred to as PATHWAYS. The project was based at Michigan State University from 1976 to 1980.

The PATHWAYS approach is based on a developmental rather than a mechanistic model of intervention for young handicapped children (Carlson, 1979). Mechanistic models have, in the past, guided intervention agents toward remedial approaches in place of developmental approaches. In remedial intervention a basic assumption is that a limited short-term experience can compensate for deficiencies acquired in prior years of development. In contrast, developmental intervention assumes competence is based on accumulated experiences and emphasizes the child's presenting state, developmental history, and the contextual environments that serve to produce intraindividual variation in behavior (see Section 3, and Bronfenbrenner, 1974). Thus, situational relativity in regard to a handicapped child's behavior becomes critically important when viewed from the more comprehensive framework of organismic theory (Carlson & Fitzgerald, 1977).

The PATHWAYS philosophy rests on the belief that *all* children are individuals and display unique and valuable differences. As with the evolutionary population approach, diversity is to be encouraged and supported in the interest of the individual and of society. Hence, education must help people develop respect for differences and expectionalities—their own and others'. Because the most appropriate time for learning this respect for self and others is when we are young, PATHWAYS project staff sought to help handicapped youngsters succesfully share nursery school and day care experiences with nonhandicapped youngsters. Our experience confirms the mutual benefit of such shared environments.

PATHWAYS staff recognized the critical importance of providing access links and supporting transitions. Hence, the PATHWAYS approach involves forming linkages with as many people and community agencies as are needed to assure the successful integration of a handicapped child into a regular nursery school or day care center (see Figure 2). As PATHWAYS personnel work on behalf of the child and family with individuals and community agencies, linkages or "pathways" are created, all of which help to make the process of mutual adaptation a successful one. For all those involved, there is an explicit recognition that successful adaptation will require *mutual* modification.

The PATHWAYS approach to helping handicapped children succeed in normal classrooms can be represented with this equation:

$$\text{Match/Fit} + \text{Support/Facilitation} = \text{Success}$$

By trying to make sure that a handicapped child is placed in a classroom suitable to his or her needs *and* by providing continuing support to children, families, teachers, cooperating agencies, and individuals, PATHWAYS seeks to ensure that everyone involved in the social and educational experience encounters as much success as possible (Carlson, 1978, 1979). Critical elements of this model will be described in the following section.

4.1. Match/Fit

The supported, integrated environment for young handicapped children is an alternative receiving much current attention as an educational intervention providing effective peer models and developmental stimulation

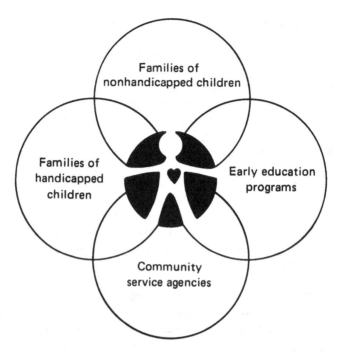

FIGURE 2. The eco-system surrounding the integrated child.

(Abelson, 1976; Bricker, 1978; Guralnik, 1976, 1978; Karnes & Teska, 1975; Martin, 1978; Meisels, 1978). Social mediation promoting acceptance and appreciation of differences without stigma is a relevant and valued aspect of integration experiences (Carlson, 1979). In order for these mediations/interventions to be successful, placement of a young handicapped child in an early childhood center must be carefully negotiated.

The placement process to be described has been created to find the most appropriate classroom environment for a young handicapped child, but the careful process of matching or creating an "adaptive fit" is, in fact, generalizable. An adaptive fit or match between the young handicapped child, the family, the teaching staff, other children, and their families is the goal. The multiple systems surrounding the child (including medicine, special education, social services, mental health, and transportation) are also considered to ensure that the probability of meeting the needs of all concerned is maximized.

Every effort was made to guarantee that the child felt as if he or she was a part of the classroom; the achievement of a positive self-concept results from successful adaptation and fosters the development of the child. The expectations that were to be placed on a child in order for that child to fully participate in classroom activities had to be within reach of the child. By the same token, it was important that the teaching staff felt able to meet the special child's needs because they were *similar* to those of other children for whom the teacher had responsibility.

The characteristics of a few of the systems impacting on the negotiation of the match/fit process will be briefly described here.

1. *The special children.* The children who were involved in this project experienced a wide variety of handicapping conditions—from deafness and other sensory and perceptual deficits to emotional difficulties, "mental retardation," and cerebral palsy. They came in varying colors, from every socioeconomic group, and were as young as eight months to as chronologically "old" as eight years. Using readily available definitions, any of the children would have been considered "moderately to severely handicapped."

The characteristics and needs of the child were very carefully examined to determine whether there was an available placement to meet these needs. Some children have great needs for specialized therapies not readily accessible; the demands placed on other young handicapped children sometimes do not allow enough energy remaining to additionally participate in an integrated setting. Some, although chronologically older, function developmentally at a very young age (below 12 months) and thus would be unable to benefit as much from the activities and structures of available classrooms, of which there are few.

Once it had been initially and carefully determined that a child could benefit from an integrated placement, an attempt was made to ascertain which center, which classroom and/or teacher(s) could best meet those needs, and what, if any, additional services or support might be necessary.

In making these decisions, the characteristics of children described in Table 1 were considered. It can be noted that many of the aspects considered before placement of a handicapped child in a regular classroom are similar to considerations given to regularly enrolling children. However, because the differences may appear to be more pronounced and may seem of more concern, they must be given special attention from the beginning. The requirement for special attention or differentiation serves to highlight the social and ecological considerations that are involved for all children (Carlson, 1983). In this way, concern with the handicap provides a focus for mutual modification and adaptation on the part of all individuals and systems involved.

2. *The special families.* Table 2 outlines the characteristics of the family system that are considered in relation to the adjunct systems (Carlson *et al.*, 1976).

3. *Early childhood centers.* It is important to solicit involvement from a variety of appropriate early childhood programs within a community to form a supportive network, but those contacts and the networking process cannot

TABLE 1. Characteristics of Special Children

Developmental level—At what developmental age does the child function? Are the majority of his or her skills at the 1-, 2-, 3-, 4-, or 5-year-old level?

Severity of handicap, safety needs, needs for attention—Will the child need an extra person in the classroom to assist in moving from one activity to the next? Is it necessary to arrange additional services, such as individual speech therapy, physical or occupational therapy?

Physical and verbal imitative skills—If a child has strong imitative skills, is it likely that an integrated setting, where appropriate and functional behavior is modeled continuously, would be at least as appropriate as a segregated setting where peer models may not be as skilled?

Absence or presence of severe aggressive or acting out behaviors—A child who manifests multiple and moderate to severe aggressive tendencies would be most appropriately placed in a setting where most of the children have learned to channel their anger. In this type of setting, children who are angry and frustrated (a natural developmental emotion) will learn nonphysical means of modifying/controlling aggression.

Needs for limits/structure—Some children have a difficult time coping with changes in their environment. For these children, the school setting might be the only predictable environment they have: the only one where limits are consistently enforced, where lunch is served at about the same time each day, and where activities follow in the same daily sequence.

Former experience with peers—The quality and types of experiences a child has previously had with children of his or her age or developmental level will offer clues regarding potential for successful integration into similar settings.

Anticipated need for adaptations—A classroom which would require a minimum number of socio-ecological adaptations is sought. However, some adaptations beyond normal individualization of instruction are expected.

Anticipated need for additional support staff—Some children may experience a time in their day when they need individual attention beyond the requirements of the rest of the children. For example, a child's poor health may not permit outside activities during the cold winter months. An additional staff person (perhaps a high school student) may stay with the child during that half hour to provide gross motor activities indoors.

Ability to separate from family—If a child does not separate easily from the family, additional support to that child (and family) may be necessary during the initial transition period. The child's transition may be assisted by a schedule of gradually increased time in the classroom supported by a significant adult in the child's life.

be described because of space limitations. However, several characteristics of the centers are noted in an effort to better match handicapped children to particular programs:

1. Type of center (day care, nursery school, and compensatory program)
2. Level of interest in participation in the integration process
3. Flexibility of program
4. Size and physical limitations of center and classrooms
5. Ratio distribution (total number of staff per class and total number of children per class)
6. Location within the community and population served

TABLE 2. Characteristics of Special Families

Family structure—Who are the members? What are the relationships? The family is the source of a child's first information about interpersonal relationships as well as his or her first experience with people. Knowing the family helps to know the child. Some of the participating classrooms have a high population of single-parent families. This setting might be comfortable for the single parent of a young handicapped child.

Family stresses—Children are very sensitive to the feelings of their parents and family. When the family is stressed, so is the child. This is as true of the "handicapped" child as it is of the "normal" child.

Interest in involvement—The extent and type of involvement that the family would like in their child's program will influence to a certain extent the type of program considered. Are the parents interested in a cooperative experience, allowing them the opportunity to be a "teacher" in the classroom? Do they need day care but want a placement that has an active parent- or child-study group with evening meetings?

Goals of parents for child (socialization and education)—The expectations that the parents have for the placement will help determine the initial placement and program. Some parents feel that their child is receiving adequate socialization at home and would like the preschool staff to concentrate on the more cognitive aspects of their child's development. Some feel that the child's education is better handled in one-to-one environments and would like the teaching staff in the integrated setting to concentrate more on enhancing social development while reinforcing the learning that is occurring in the individualized setting.

Need for relief from responsibilities of the handicapped child—The families' need for relief may expedite the placement process. In some cases, placement in a day care setting is necessary in order for the family to continue functioning.

Availability of family support networks; availability of other support structures and services— If a family does not have adequate support within the primary system, other support services are sought. Some families have already located support structures. Others need assistance in doing so (i.e., referral to the Family Life Clinic or the Department of Social Services).

Need for tuition/financial concerns—Primary survival needs for food and shelter must be met before other needs can be dealt with. If a family is experiencing financial difficulties, tuition for the child may be provided and additional assistance will be sought.

Transportation needs—All too often, lack of transportation has been the stumbling block in allowing a child and his or her family to participate in a program. In cooperation with school districts, the Department of Social Services, and regularly enrolled families, transportation problems will be resolved.

7. Staff qualifications, attitudes, and related information
8. Center and staff stability
9. Program philosophy
10. Program content
11. Potential for providing socialization experiences
12. Language enrichment experiences
13. Positive alternative role modeling
14. Description of the children typically enrolled in the center (age and special needs)

15. Center's (and/or staff's) former experience with handicapped children
16. Center's (and/or staff's) contacts with support agencies/people
17. Existing parent program

With at least this information in mind, PATHWAYS staff met with each child and the child's family or caregiver(s). The pattern was for an individual to be assigned to a prospective family when the child had been referred to the project. This individual became the case facilitator and mediated the transition and continual functioning of the child and family within the program. The case facilitator (a) assessed the child's integration needs in conjunction with the child's family or caregiver(s), (b) with others recommended a placement, and (c) throughout the integration process, provided support to the child, family, and school. The case-facilitator role was designed to better meet the needs of each family by providing one supportive contact person and thereby decreasing many of the contacts usually necessary with other professionals in the case.

The next part of the process involves reciprocal observation and discussions. Home visits, individual contacts, tentative suggestions, checklists of behavior, goal setting, placement options, and center visitations were considered/provided in order for everyone to feel as informed as possible before the initial placement decision was made. Since voluntary participation has always been an operational guideline, all parties—the family, the teachers, and others concerned—sought to find an acceptable placement recommendation.

When everyone agreed to a placement suggestion transportation was arranged and the child began attending the nursery school or day care center classes on a provisional basis. When the child entered the classroom, an Individualized Educational Program was prepared and begun. The child and the classroom environment were closely observed to assess the success of the child in adapting to the new environment and to more fully develop the Individualized Educational Program required by current law. Because each child in the PATHWAYS program was unique, each followed a different educational plan.

Each child and each classroom also followed a unique course of transition and adjustment. For some children, adapting to the new classroom environment was a relatively easy or short-lived task; for others it was a more difficult or lengthy experience. Adaptational differences were also found in teachers and in families. Anecdotal records were used by the case facilitator to help evaluate the success of the placement and the child's progress, as well as to provide a comprehensive data base for future analysis. In preparing records and in analyzing how well a child was adapting to the new environment and how well the environment was adapting to the child, the following ethnographic information was considered:

- How the separation between parent and child was accomplished
- The relationship between the handicapped child and the other children
- The adult-child relationships that take place
- How the child interacts during a structured group experience
- How the child moves about the classroom
- How much initiative the child shows in dealing with people and materials in the classroom
- How the child responds to classroom elements such as music, nonsense and humor, disruptions or changes, instructions and direction, and discipline

If it seemed that the adaptive fit was *not* evolving, as in the case of Jason, the initial placement was reassessed by the case facilitator, and a potentially more appropriate placement was considered. If the placement seemed successful, a revised Individualized Educational Program was prepared, based on an evaluation of the child's progress. The planning of this revised program was done jointly by the case facilitator, the child's family, the classroom teacher, and other interested individuals.

4.2. Support/Facilitation

The second part of the PATHWAYS equation for success is support/facilitation. The successful integration of young handicapped children into existing early childhood programs requires the participation and cooperation of many different individuals. For some of those involved, this may be an unfamiliar experience, requiring contact with members of the community outside of known systems of contact. As a support system (technical assistance) model, PATHWAYS staff anticipated and responded to the needs of key groups of individuals: children, their parents, their teachers, and participating agencies. Together, these people comprise an "eco-system" that surrounds the individual integrated child (see Figure 2). The rationale encompassing the support services offered to these different groups is based on fundamental observations of what individuals need in order to become involved in change and development. Consideration centered on the expectations that people bring to participation in a social or an educational innovation. We have observed the following:

1. Most individuals have initial concerns when faced with a new experience. They may feel threatened, may be concerned with skill

inadequacy, or worried that they stand to lose services, status, or power.

2. Every person involved is unique and has individual needs, concerns, and abilities. Each person has his or her own way of learning, receiving information, and using resources.

3. When individuals are offered alternatives and choices and are encouraged to define their own needs, they are able to be active participants in a process and better able to generate and implement their own solutions. In short, they are better able to define their own developmental contexts.

4. Transitions (psychological or physical movement from one situation to another) are a challenge to most individuals and it is during these times that most services and support will be needed.

5. Willingness to participate in a new experience is highly important for success. Individuals who voluntarily elect to participate will be more likely to become partners in the process.

6. Ongoing services, periodically reassessed for effectiveness, must be adequately and and continually clarified.

7. Ongoing communication among participants is extremely valuable and communication networks should be encouraged.

8. All individuals need recognition and positive acclaim for their accomplishments. Development is facilitated by a mutually beneficial feedback process.

An understanding of these complex needs forms the basis of the support/ facilitation component of the PATHWAYS model. We have found that when individuals feel that their needs and abilities are considered and respected, they are better able to work toward a common goal of providing the most appropriate early educational experiences for young handicapped children (Carlson, 1979).

As a technical assistance model, PATHWAYS provides services in order to implement and evaluate the carefully negotiated placement of an individual handicapped child. These services have two primary functions: they are supportive of the individuals involved, and they are facilitative of further growth and development. PATHWAYS services to children, teachers, parents, and agencies are supportive; helping to provide a needed security while developing new competencies and while coping with ongoing responsibilities. Each individual has unique needs; support services vary according to those needs. The combined effect is to create a positive climate for ongoing educational experiences of children with diverse needs and abilities. The support PATHWAYS provides tends to fall into three categories: family support, educational program support, and interagency coordinative support.

4.2.1. Family Support

Family support is given to members of the primary system throughout. The type of support provided the members ranged from helping to find a way to transport the child to and from school to developing linkages between families for mutual help and problem-solving. It was the case-facilitator's responsibility to work with each family and assess the needs of this system in building an adjunctive support system. The case facilitator tried to make sure that many services were provided: human support, educational services, physical and financial help.

4.2.2. Educational Support

Educational support was given to the center and teacher when needed to help them better accommodate the presence of a handicapped child. Many early childhood education teachers have had little experience with handicapped children. They may also have had little previous information about or training in the education of children with special needs. Consequently, they may want to discuss any anxieties they have about having a handicapped child in their classroom. Alternatively, they may desire information about structuring the environment so as to provide the best possible development context for the child (see Carlson & Wandschneider, 1979). It was the task of the case facilitator to attempt to help the teacher develop confidence and skill in adapting the environment and in helping the handicapped child adapt.

4.2.3. Interagency Support

Interagency support was a result of PATHWAYS staff efforts to coordinate activities that affect the child. Through face-to-face contacts, telephone calls, and a newsletter, PATHWAYS informed other agencies involved with the families about opportunities offered to participating children. The staff also sought consultation services and made referrals to agencies that might help the child or the child's family in the future.

As the equation stipulates, success was and is the outcome of this process. Success, however, was measured not just in terms of the achievement of an adaptive fit for the young handicapped child and family, but also for the nonhandicapped children in the classroom and their families, the teacher, director and staff of the center in which the child was placed, and the agencies and individuals involved in the adjunct systems of which the family was a part.

4.3. Research Dimensions

The socio-ecological model just described was based on several assumptions (many of which have been briefly detailed throughout this chapter). In this project, however, a comprehensive research/evaluation component was added to explore several basic research questions, as well as to look at the underlying assumptions. The results of this research may be found in the PATHWAYS Final Report (Carlson, 1979). That the integration of young handicapped children can be successful within a particular framework is documented on 27 pages and 26 tables (pp. 77–104).

What may be of more interest to the reader, however, are the particular research questions that were asked. If one were to implement a socio-ecological perspective in behavioral pediatrics, questions similar to these might be used. In this project the research questions were divided into two main categories: feasibility and effectiveness. It appeared to the investigating team that effectiveness could not honestly be analyzed without first determining under what socio-ecological conditions the scope and demands were feasible.

4.3.1. Feasibility

- What are the regular classroom teacher time requirements for integrating young handicapped children? (See Carlson, Renshaw, & Andrews, 1981)
- What are the impacts on the other regularly enrolled children?
- What are the impacts on the classroom support staff?
- What types of support services do classroom teachers need? (See Carlson & Wandschneider, 1979; Spell & Carlson, 1981).
- What are the specific and overall reactions of teachers to participation in the project?
- What personal qualities, skills, and knowledge are necessary in an individual in order to successfully implement the case-facilitator role?
- How do professionals, parents, and other members of the community view the feasibility of integration in early childhood?
- How many early childhood centers in the local, regional and state areas are willing to negotiate in regard to integrating young handicapped children?
- What support was needed by PATHWAYS staff? In what ways did it differ from other supportive processes?

4.3.2. Effectiveness

- What is the average change in rate of development for the integrated handicapped child compared with what could be expected if no intervention occurred? (Motor, communication, cognitive, and social development measures)
- What is the handicapped child's level of social interaction in the classroom compared to typical peers in the same classroom?
- What were transition periods like for children, families, teaching staff, PATHWAYS staff, others? What types of assistance were helpful?
- What were the changes in functional skills and competencies of teachers?
- How did PATHWAYS teachers compare to other early childhood teachers (integrating and nonintegrating) in terms of attitude, knowledge, and experience? (See Clapp, 1979.)
- Were there changes in teacher's feelings of confidence about working with a handicapped child? What types of changes?
- Were there differences in the amount of willingness to work with handicapped children in the future?
- How satisfied were parents of the regularly enrolled and the special children with their child's experience in an integrated classroom?
- Were parents willing to enroll their children in an integrated classroom in the future?
- How much did the parents of handicapped children increase in skill and confidence in working with their child? In working with agency personnel?
- What forms of communication in the network were most effective? Least?

Although these questions do not represent all that were asked (or indeed, all the data forthcoming), they do illustrate the scope necessary to begin to analyze an applied program based on a socio-ecological perspective.

The information regarding PATHWAYS has been included in this chapter for illustrative purposes. To many people, PATHWAYS represents a modestly successful attempt to grapple with the real-life complexities of adaptation, given a number of different, continuously changing environments. The concepts represented in the socio-ecological theory of adaptive behavior and functioning were mapped early on in the project. It was possible to study, at least in a limited way, the following elements of the model: (a) the presenting state of the young child considered to be "handicapped"; (b) the characteristics of the varying socio-ecological environments in which the child was located, including but not limited to, the primary system/family, the early childhood classroom, and any other relevant environments (special education classroom, special therapy situations, hospital or clinic, and diagnostic facilities); (c) exchange processes between the child and aspects of the environment, for example, cognitive and social interactions with teachers, other children, and family; (d) transitions within and across the varying environments that included attention to transportation agents and the transportation system itself, as well as transitions into and out of the center and transitions between daily activities; and (e) the negotiation of an adaptive fit or match between the child and the varying environments, with particular emphasis on the early childhood classroom.

Since success was stipulated in the equation, and clarified for all the volunteer participants, it was expected that the process would be successful. What was not expected was the degree to which this process was successful. The key may be the expectation and challenge of mutual modification. There were no "right/wrong" dimensions to the process; only continual negotiation until a temporary, but satisfactory, adaptation was achieved for the individuals involved.

We believe that the theory, process, and outcome are generalizable; that is, there are issues of equal or more importance in behavioral pediatrics today that could be approached from a socio-ecological perspective. Attention to the relevant dimensions may yield surprising but satisfying results. At the very least, we might suggest failure to thrive syndrome; sudden infant death syndrome; child abuse and neglect; teen-age pregnancies; enuresis in middle and later childhood; juvenile diabetes; anorexia nervosa in adolescence; Reyes Syndrome; and sex education. Doubtless the reader has a research or clinical interest that could easily be included.

A critical element in clinical work is an appreciation of continually negotiated and supported processes of mutual adaptation. Adaptation of family, medical, and educational systems, with an anticipated outcome of and value for adaptive coherence, has not been the rule in our contemporary society.

5. BEHAVIORAL PEDIATRICS: PARADIGMS FOR AN EMERGING DISCIPLINE

Because the domain of behavioral pediatrics is, at present, a young and emerging discipline; a variety of research paradigms are being offered for its future guidance. The material in this chapter is presented as an effort to articulate and demonstrate the characteristics and research/programmatic implications of one particular paradigm. It is a paradigm that is itself in the process of development and application.

We have suggested that behavioral pediatrics be construed as a most wide-ranging and encompassing discipline. It is a discipline that focuses on developing persons and their behavior, and the behavior of those with whom they interact. Our analysis suggests that to this concern may be added a concern with both the primary and adjunct environments in which behavior is situated. From this derives the additional concern with linkages between environments, and the manner in which cross-environment transitions are negotiated.

The process of development is viewed as a necessarily situated process in which the organism increases in complexity. The increase in complexity is mediated by the organism's exchange processes with its environment. In that development is a historical process involving transactions with a variety of environments, the notion of a developmental trajectory has been employed to capture and unify this perspective.

The socially and ecologically suited character of development suggests that a central emphasis be placed on the process of adaptation. In addressing the process of adaptation, a combined evolutionary and developmental perspective has been employed. With this perspective comes an emphasis on individual uniqueness and population diversity. A correlated emphasis is placed on processes of mutual modification and the negotiated achievement of an adaptive fit.

Beyond its general import for an understanding of development and behavior, it is suggested that the socio-ecological paradigm provides a framework for research on health care environments, behavior, and outcomes (Cassel, Hogan, & Gallagher, 1979). By focusing attention on the environments in which health care activities are carried out, and the interactive behaviors that mediate these activities, new research questions may be brought to light. At the same time, new techniques for influencinghealth care outcomes, such as environmental modifications, may be considered.

The disciplined investigation of the social and ecological bases of human behavior and adaptation will result in basic and in applied information. Whether the pediatrician is concerned with gaining information for anticipatory guidance, with understanding the relation of the therapeutic alliance to the achievement of healthy functioning, with reorganizing trianing environments, or with a myriad of other problems, we believe that a socio-ecological perspective will be of value. We look forward to the development and application of this life-sciences paradigm to additional research and theory in behavioral pediatrics.

6. REFERENCES

Abelson, A. G. Measuring preschools' readiness to mainstream handicapped children. *Child Welfare*, 1976, *55*, (3), 216–220.

Aschoff, J., Fatranska, M., Giedke, H., Doerr, P., Stamm, D., & Wisser, H. Human circadian rhythms in continuous darkness: Entrainment by social cues. *Science*, 1971, *171*, 213–215.

Bohm, D. *The special theory of relativity.* New York: W. A. Benjamin, 1965.

Bonner, J. T. *On development: The biology of form.* Cambridge: Harvard University Press, 1974.

Brazelton, T. B., Koslowski, B., & Main, M. The origins of reciprocity: The early mother-infant interaction. In M. Lewis & L. Rosenblum (Eds.), *The effect of the infant on its caregiver.* New York: 1974.

Brent, R. L., & Morse, H. B. Not in our own image: Educating the pediatrician for practice. *Pediatric Clinics of North America*, 1969, *16*, 793–808.

Bricker, D. D. A rationale for the integration of handicapped and nonhandicapped preschool children. In M. J. Guralnick (Ed.), *Early intervention and the integration of handicapped and nonhandicapped children.* Baltimore: University Park Press, 1978.

Bronfenbrenner, U. *Is early intervention effective? A report on longitudinal evaluations of preschool programs* (Vol. 2). Washington, D.C.: Office of Child Development, 1974.

Carlson, N. A. (Ed.). *The contexts of life: A socio-ecological model of adaptive behavior and functioning* (Final report to Bureau of Education for the Handicapped, USOE/BEH). East Lansing, Mich.: Institute for Family and Child Study, Michigan State University, 1976.

Carlson, N. A. (Ed.). *Program narrative, PATHWAYS project: A human support system model for integrated handicapped children and their families.* East Lansing, Mich.: Institute for Family and Child Study, Michigan State University, 1978.

Carlson, N. A. (Ed.). *PATHWAYS; Toward integrating handicapped children* (Final report to Bureau of Education for the Handicapped, USOE/BEH). East Lansing, Mich.: Institute for Family and Child Study, Michigan State University, 1979.

Carlson, N. A. Toward integrating handicapped children: A case study analysis of the educational context and the least restrictive environment. *Technology and Handicapped People* (Working Paper no. 2). Washington, D.C.: U.S. Congress, Office of Technology Assessment, January 1983.

Carlson, N. A., & Fitzgerald, H. E. *A proposal for the establishment of an Institute for the Study of Early Identification for the Handicapped* (Proposal submitted to Bureau of Education for the Handicapped, USOE/HEW, Washington, D.C.). East Lansing, Mich.: Institute for Family and Child Study, Michigan State University, 1977.

Carlson, N. A., & Wandschneider, M. R. (Eds.). *Toward integrating your handicapped children: A handbook for preschool teachers, parents and referring agencies.* East Lansing, Mich.: Institute for Family and Child Study, Michigan State University, 1979.

Carlson, N. A., Lafkas, C., & Cassel, T. Z. Some applications of the socio-ecological model of adaptive behavior and functioning. In N. A. Carlson (Ed.), *The contexts of life: A socio-ecological model of adaptive behavior and functioning* (Final report to Bureau of Education for the Handicapped, USOE/BEH). East Lansing, Mich.: Institute for Family and Child Study, Michigan State University, 1976.

Carlson, N. A., Renshaw, L., & Andrews, M. Teacher perceptions of time requirements for integrating young handicapped children. *The Exceptional Child*, 1981, *28*(2), 119–127.

Cassel, T. Z. *The linkage between the structural complexity of the individual and that of the context.* Paper presented at the Conference on Cognitive Development, Boston, 1972.

Cassel, T. Z. *The context of development: Four aspects of a general model and a note of the functions of attachment.* Paper presented at the Seminar on Development Issues, Boston, 1974.

Cassel, T. Z. A socio-ecological model of adaptive functioning: A contextual developmental perspective. In N. A. Carlson (Ed.), *The contexts of life: A socio-ecological model of adaptive behavior and functioning* (Final report to Bureau of Education for the Handicapped, USOE/ BEH). East Lansing, Mich.: Institute for Family and Child Study, Michigan State University, 1976.

Cassel, T. Z. *Research in primary care pediatrics: Environmental mapping and longitudinal studies.* Paper prepared for the Department of Pediatrics, Wayne State University School of Medicine, 1978.

Cassel, T. Z., & Sander, L. S. *Neonatal recognition processes and attachment: The masking experiment.* Paper presented at the meeting of the Society for Research in Child Development, Denver, 1975.

Cassel, T. Z., Hogan, M. J., & Gallagher, R. E. *The emergence of medical behavioral ecology.* Paper prepared for the Division of Educational Services and Research, Wayne State University School of Medicine, 1979.

Clapp, G. M. *A comparative study of the attitudes, knowledge and experience of teachers regarding handicapping conditions and mainstreaming in early childhood programs.* Unpublished doctoral dissertation, Michigan State University, 1979.

Clutton-Brock, T. H., & Harvey, P. H. *Primate ecology and social organization. Journal of Zoology* (London), 1977, *183*, 1–39.

Condon, W. S., & Sander, L. W. *Neonatal movement is synchronized with adult speech: Interactional participation and language acquisition. Science,* 1974, *183*, 99–101.

Crook, J. H. The adaptive significance of avian social organizations. *Symposia of the Zoological Society of London,* 1965, *14*, 182–218.

Crook, J. H. The socio-ecology of primates. In J. H. Crook (Ed.), *Social behavior in birds and mammals.* London: Academic Press, 1970.

Cullen, E. Adaptations in kittiwake to cliff nesting. *Ibis,* 1957, *99*, 275–302.

Dewey, J. *Logic: The theory of inquiry.* New York: Holt, Rinehart & Winston, 1938.

Eisenberg, J. F., Muckenhirn, N. A., & Rudran, R. *The relation between ecology and social structure in primates. Science,* 1972, *176*, 863–874.

Guralnick, M. J. The value of integrating handicapped and nonhandicapped preschool children. *American Journal of Orthopsychiatry,* 1976, *46*, 236–245.

Guralnick, M. J. (Ed.). *Early intervention and the integration of handicapped and nonhandicapped children.* Baltimore: University Park Press, 1978.

Hinde, R. A., & Stevenson-Hinde, J. (Eds.). *Constraints on learning: Limitations and predispositions.* New York: Academic Press, 1973.

Huxley, J. S. Courtship activities in the red-throated diver (*Colymbus Stellatus Pontopp*) together with a discussion of the evolution of courtship in birds. *Journal of Linnaeus Society,* 1923, *35*, 253–292.

Karnes, M. B., & Teska, J. A. Children's response to integration programs. In J. J. Gallagher (Ed.), *The application of child development research to exceptional children.* Reston, Va.: Council for Exceptional Children, 1975.

Klaus, M. H., & Kennell, J. H. *Maternal-infant bonding: The impact of early separation or loss on family development.* St. Louis: C V Mosby, 1976.

Kuo, Z. Y. *The dynamics of behavior development.* New York: Random House, 1967.

Martin, E. W. Some thoughts on mainstreaming. In G. J. Warfield (Ed.), *Mainstream currents.* Reston, Va.: The Council for Exceptional Children, 1974.

Meisels, S. J. First steps in mainstreaming: Some questions and answers. *Young Children,* 1977, *33*(1), 4–13.

Sander, L. W. Issues in early mother-child interaction. *Journal of the American Academy of Child Psychiatry,* 1962, *1*, 141–166.

Sander, L. W. Adaptive relationships in early mother-child interaction. *Journal of the American Academy of Child Psychiatry*, 1964, *3*, 231–264.

Sander, L. W. The longitudinal course of early mother-child interaction: Cross-case comparison in a sample of mother-child pairs. In B. M. Foss (Ed.), *Determinants of infant behavior*. London: Methuen, 1969.

Sander, L. W. Infant and caretaking environment: Investigation and conceptualization of adaptive behavior in a system of increasing complexity. In E. J. Anthony (Ed.), *Explorations in child psychiatry*. New York: Plenum Press, 1975.

Schaffer, H. R. (Ed.). *Studies in mother-infant interaction*. New York: Academic Press, 1977.

Spell, J., & Carlson, N. Mainstreaming at the preschool level? *Early Years*, May 1981, pp. 70–71.

Stern, D. N. The goal and structure of mother-infant play. *Journal of the American Academy of Child Psychiatry*, 1974, *13*, 402–421.

Sullivan, H. S. *Conceptions of modern psychiatry*. New York: Norton, 1953.

Tobach, E. Aronson, L. R., & Shaw, E. *The biopsychology of development*. New York: Academic Press, 1971.

Wolpe, J. *The practice of behavior therapy*. New York: Pergamon Press, 1973.

3

Neural Plasticity
A New Frontier for Infant Development

FRANK H. DUFFY, GEORGE MOWER,
FRANCES JENSEN, AND HEIDELISE ALS

1. INTRODUCTION

It is a common observation in infant follow-up studies that those children
with histories of premature birth (premies) manifest different behavioral
profiles from those born at term. Even if major medical complications are
excluded, one hears such adjectives applied as "rigid," "static," "sober,"
"irritable," "poorly modulated." Drillien, Thomson, and Bargoyne (1980)
notes that up to 50% of small for gestational age (SGA) premies have some
degree of learning disability. It has become increasingly clear that not everything can be explained as complications of known cerebral insult (anoxia,
hemorrhage, trauma, and seizures). Accordingly, investigators (Als, 1983;
Rose, 1981) have come to wonder if the simple fact of premature birth may,
in and of itself, have important consequences for subsequent neural development. Should this new perspective prove correct, then management of the
premies' environment may take on an importance equal to the management
and the prevention of medical complications.

The clinical relevance of this environmentally oriented developmental
perspective rests on three assumptions. *First,* there must be clinical evidence
to suggest that "uncomplicated" premature birth results in altered, less than
optimal outcome, which appears to be a valid although not yet closely documented assumption. As previously mentioned, premies are burdened with

Frank H. Duffy, George Mower, and Frances Jensen • Developmental Neurophysiology
Laboratory, Seizure Unit and Department of Neurology, Children's Hospital Medical Center
and Harvard Medical School, Boston, Massachusetts 02138. **Heidelise Als** • Child Development Unit, Children's Hospital Medical Center and Harvard Medical School, Boston, Massachusetts 02138.

a higher incidence of subtle behavioral and neurological differences; however, the relative roles of medical illness and of environmental influence have not yet been established.

Second, adequate behavioral, neurological, and/or neurophysiological assessment methodologies must exist or be developed to document differential effects of the environment. Fortunately, there is recent work in this area with the development of increasingly subtle behavioral and neurophysiological means to investigate such issues (Als, Lester, Tronick, & Brazelton, 1982a,b; Amiel-Tison, Barrier, Shnider, Levinson, Hughes, & Stefani,1982; Brazelton, 1973; Daum, Grellong, Kurtzberg, & Vaughn, 1977; Dubowitz & Dubowitz, 1981; Duffy,, 1982; Duffy, Burchfiel, & Lombroso, 1979; Lombroso, 1975, 1979). Preliminary evidence, from our own laboratory, suggests that the differences in behavioral organization of neonates are associated with neurophysiological differences overlying frontal lobe (Duffy & Als, 1983). This finding is of interest as our study of older children with learning disabilities also demonstrates prominent frontal lobe difference (Duffy, Denckla, Bartels, & Sandini, 1980). Although it is compelling, the thread of evidence from this laboratory and others is thin and awaits further documentation.

Third, the animal laboratory must provide compelling evidence that environmental modification has an impact on the development of the central nervous system. Moreover, to be clinically relevant to the issue of prematurity, environmental manipulations must be of the same order of magnitude as the differential that exists between the human intra- and extrauterine environment. We shall devote the remainder of this chapter to a review of the relevant data as they pertain to the development of the normal animal and of the brain-lesioned animal. As we shall see, there is strong support for a critical environmental influence on normal brain development and on recovery from brain lesions.

2. EFFECT OF ENVIRONMENTAL MANIPULATIONS ON BRAIN DEVELOPMENT OF NORMAL, UNLESIONED ANIMALS

One might explore many research avenues to document environmental impact on neural development. For example, abundant evidence exists to suggest that adult birdsong is molded by early experience (Baker, Spitler-Nabors, & Bradley, 1981; Marler & Peters, 1981; see review by Marx, 1982). Moreover, studies of the remarkable cyclical fluctuations of the adult canary song control centers may provide, among other fascinating possibilities, a model for the study of the brain's potential for repair (Nottebohm, 1981). However, the best documented and most clinically relevant research appears to involve the cat and the monkey visual systems. Of particular importance has been the pioneering work of Hubel and Wiesel (1959, 1960a,b, 1962, 1963, 1965a, 1968) as recently reviewed by Barlow (1982). Visual pathways are prominent and are easily studied by neurophysiological and neuroanatomical techniques. Moreover, paradigms to modify visual experience are readily developed. Despite this emphasis on the visual system, it is likely that much of the work can also be generalized to other systems: motor, auditory, and somatosensory.

2.1. Environmental Effects on Visual Development

2.1.1. Visual System Basics

There are three important way stations in the flow of visual information from the eye to the visual cortex of all species (see Figure 1). First, there is the retinal ganglion (RG) cell output neuron of the eye, whose axons form the optic nerves. Next, there is the lateral geniculate nucleus (LGN), the thalamic relay nucleus deep within the hemisphere, which receives RG cell input, and whose axons form the projections to visual cortex (VC). Finally, there is the VC itself, encompassing three functionally and anatomically separate regions: areas 17, 18, and 19. Connections from RG cells to the LGN neurons, from LGN to VC, and among areas 17, 18, and 19 of VC are highly specific and ordered in the adult animal. Within LGN, cells are exclusively monocular, being activated by stimulation of just one eye. Cells are not randomly distributed within each LGN but are arranged in layers of alternating monocular influence: three layers in the cat and six in the monkey. Binocular interaction first occurs within VC. However, not all cells are equally activated by both eyes. A simple way to describe this relative responsiveness to stimulation of the two eyes is to classify single VC cells into one of five categories. Categories 1 and 5 comprise cells activated exclusively by one or the other eye. Categories 2 and 4 consist of cells dominated by one or the other eye but with binocular influence. In Category 3, cells are equally responsive to each eye. By sampling large numbers of cells, an ocular dominance (OD) distribution plot is generated that provides an overall description of the relative influence of each eye (see Figure 2, Section A).

Recent evidence demonstrates that ocular dominance is not randomly distributed throughout VC. Since the first physiological reports of aggregations by ocular dominance of cells in monkey VC (Hubel & Wiesel, 1965a, 1968), numerous neuroanatomical mapping procedures have been used to confirm the existence and to define the nature of ocular dominance columns (Hubel & Freeman, 1977; Levay, Wiesel & Hubel, 1980). For example, if one eye of a monkey is injected with a radioactive amino acid (e.g., tritiated

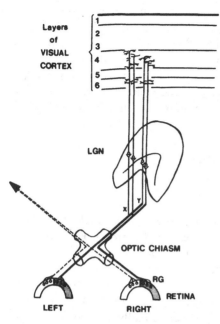

FIGURE 1. Organization of the visual system. This figure illustrates in a simplified and diagrammatic manner the visual system of the cat and the primate. Pictured below as semicircles are the retinas of the two eyes. The right half of each retina, which receives input from the left external visual field, is stippled. Within each retina the small circles represent the retinal ganglion (RG) cells that give rise to the axons of the optic nerve. As can be seen, RG cell axons project back to the optic chiasm where they alter course for the main thalamic relay station of the visual system, the lateral geniculate nucleus (LGN). Axons from the right hemiretina project to the right LGN. Within LGN, however, RG axons from each eye remain segregated into independent layers, the number of which is species dependent. A synapse occurs in the LGN with output axons coursing primarily to the visual cortex (VC) via the optic radiations. The right retina projects exclusively to the right LGN and VC. Thus, visual information from the left-visual field impinges on the right retina and is exclusively processed within the right-cerebral hemisphere. The VC has six layers with most LGN fibres synapsing in layer IV. Within VC, many cells receive input from both eyes. Thus, binocularity of the visual system first occurs at the cortical level. The VC has at least 3 histologically recognized subdivisions known as area 17, 18, and 19. To some degree there are at least two parallel types of information that pass to VC from the eyes. The X system has slow conduction velocity but high spatial-resolving power, primarily subserves central vision, projects to VC area 17, and is believed important for fine detail discrimination. The Y system has faster conduction velocity but lower spatial resolving power, primarily subseves peripheral vision, projects to VC areas 17 and 18, and is believed important for the detection of moving objects, especially in the periphery. We have not included diagrammatic representations of the W system or the projection of the visual system to other portions of the visual system such as the superior colliculus. A broader review of this connectivity may be found in the review of Stone and Dreher (1982).

proline), the substance is transported to VC transneuronally. By autoradiography, a pattern of alternately labelled bands or stripes appears (see Figure 3, Section A). Labelled areas represent the terminal zones of afferent axons from LGN connected to the injected eye. Overall, in the normal adult animal, equal areas of cortex are influenced by each eye and regions of influence are largely segregated into alternating cortical stripes, each approximately 400 microns wide.

Individual cells at all levels within the visual system respond to specific retinal stimulation. The term *receptive field* (RF) refers to the area of retina that, when stimulated, causes a cell to respond. Moreover, the nature of stimulation within the RF may differentially effect cell firing. To map RFs,

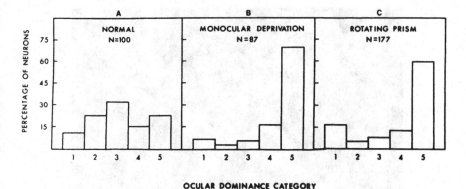

OCULAR DOMINANCE CATEGORY

FIGURE 2. Ocular dominance distribution. Binocular inputs to single neurons are first appreciated at the level of the VC. By observing the electrical response of VC neurons to separate stimulation of each eye, one not only demonstrates binocularity *per se*, but one also observes different degrees of influence from each eye on each particular neuron. The relative influence of each eye on a cortical neuron is referred to as ocular dominance (OD). Numerical descriptors of ocular dominance have been developed along a scale of 1 to 7 or, as shown in this figure, from 1 to 5. Along the 1 to 5 axis, cells in OD category 3 demonstrate equal influence from each eye. Those in category 1 are exclusively influenced by the eye on the opposite (contralateral) side and those in category 5 are influenced exclusively by the eye on the same (ipsilateral) side. Cells in categories 2 and 4 demonstrate binocular inputs, but show asymmetrical influence from the eyes. Section A shows the OD histogram of normal kittens with somewhat quantitatively limited but qualitatively normal binocular visual experience during the critical period. Most cells are binocular (categories 2, 3, and 4). In kittens with completely normal experience, the monocular cells in categories 1 and 5 would be even further reduced in frequency (not illustrated). Kittens reared with bilateral pattern vision deprivation from bilateral lid suture (BS) during the critical period show a loss of binocular cells of category 3 and a bimodal distribution of increased frequency in monocular categories 1 and 5 (not illustrated). Section B shows the OD histogram for kittens raised in the monocular deprivation (MD) paradigm during their critical periods. Note the almost complete loss of binocularity (categories 2, 3, and 4) and of response to the deprived eye (in this case, category 1). Section C shows the kitten OD histogram resulting from our monocular rotating prism (MRP) paradigm (see text). The kittens experienced normal amounts of light and pattern impinging on each retina. However, the image to one eye had been rendered unreliable in its relationship to external space by causing it to be passed through a wedge prism which was randomly and frequently rotated to a new position. The effect of the MRP on the OD histogram is similar to that produced by MD, that is, loss of binocularity and suppression of cortical influence from the treated (MRP) eye (in this case, category 1).

animals are immobilized and their eyes are made to focus on a photographic projection screen. Thus, the retina is topographically mapped on this screen and RFs are determined by projecting visual stimuli at differing screen positions. The response of a single cell to retinal stimulation is determined by recording electrical activity with tiny, sharpened electrodes insulated up to the tip. Using this extracellular microelectrode recording process, Hubel & Wiesel (1962, 1963, 1965a) suggested that RF organization within the visual system of cat and monkey is highly organized and hierarchical in nature (see

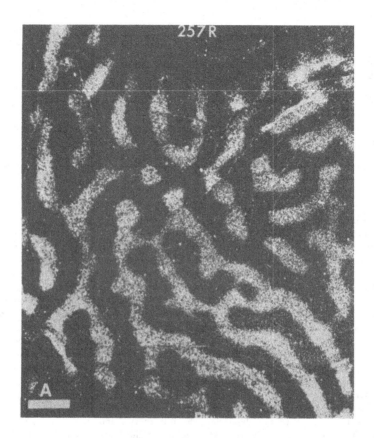

FIGURE 3. Ocular dominance autoradiograms. VC neurons from the different OD categories (see text and Figure 2) are not randomly scattered throughout cortex. Cells dominated by one eye (categories 1 and 2) are grouped separately from those dominated by the other eye (categories 4 and 5). This can be graphically demonstrated by the intraocular injection of radioactive tritiated (radioactive) proline that is transneuronally transported from the eye to VC. Subsequent auto radiograms reveal the distribution of VC neurons primarily dominated by LGN fibres from the injected eye. In Section A we see the results of a section of VC cut tangentially to layer IV. LGN axon terminals appear white, forming the light stripes or bands corresponding to the injected eye. Note that the light (injected eye) and dark (normal eye) stripes are of equal size for this 6-week-old normal monkey. The same pattern is seen in the normal adult monkey (not illustrated). Section B is a similar autoradiogram of a MD monkey whose right eye was closed at two weeks of age and whose left eye was injected 18 months later. Note the enlarged size of light stripes dominated by the normal eye and diminished size of dark stripes dominated by the deprived eye. Section C is yet another autoradiogram from another MD monkey whose eyes were closed at three weeks and whose deprived eye was injected at 6 months of age. Note the same findings as Section B, but in Section C the deprived eye was injected. From *Journal of Comparative Neurology*, 1980, *191*, 1–51. Copyright 1980 by Alan R. Liss, Inc. Reduced respectively at 80%, 80%, and 71%. Reprinted by permission of the author and the publisher.

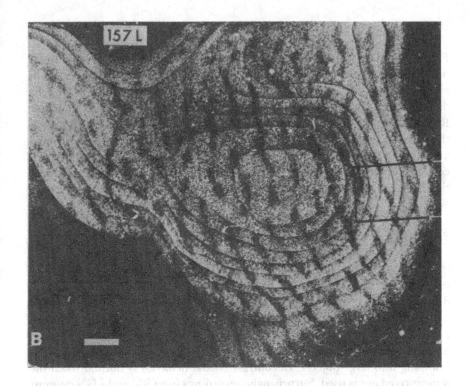

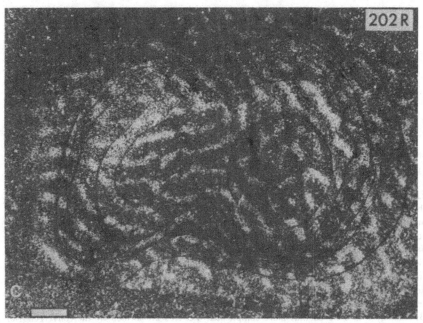

Figure 4). RG and LGN neurons prefer small stationary spots of light and fire to the onset and/or offset of light. In VC, cells tend to prefer linear stimuli either stationary or moving. It is believed that the projection of LGN cells on VC neurons is not random but, in fact, ordered so that the small circular RFs of LGN line up to form the linear RFs of VC (see Figure 4). Moreover, the more complex RFs seen in VC areas 18 and 19 are believed to be formed as the summation of inputs from cells with more simple RFs. Within VC, cells responding to binocular stimulation almost always demonstrate identical RFs to stimulation of each eye independently.

2.1.2. Monocular Deprivation Amblyopia and Similar Paradigms.

It was Wiesel and Hubel (1963a,b, 1965) who opened up the field of plasticity by investigating the neural and the behavioral consequences of depriving kittens of monocular visual experience from time of birth. Profound behavioral and electrophysiological alterations are noted when the deprived eye is subsequently opened months later and individually tested. The cat appears blind in the monocularly deprived (MD) eye, barely able to avoid bumping into large objects. In contrast, visual behavior is normal when the nondeprived eye is used. Surprisingly, recordings from RG and LGN neurons

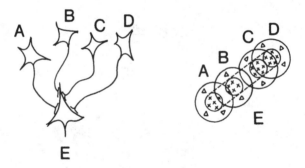

FIGURE 4. Hierarchical organization of receptive fields. Hubel and Wiesel (1962, 1965a) were the first to suggest that the linear receptive fields (RFs) of VC result from the specific projection onto one VC neuron of LGN axons whose individual RFs contribute to the new RF of the cortical cell. This is schematically illustrated where A,B,C, and D are LGN cells projecting onto a single VC neuron, E. To the right, the RF of E is shown to be a linear region of excitation (within dotted lines) formed as the summation of small circular RFs of the four LGN cells. The small LGN excitatory centers are shown as circles filled with plus signs. The LGN RFs also demonstrate surrounding annular inhibitory regions illustrated by small triangles. The VC receptive field has a long, slanted central area of excitation and surrounding regions of inhibition. By analogy it is believed that more complicated VC RFs are formed by the convergence of inputs from VC neurons with simpler RFs.

driven by the MD eye reveal only subtle differences from normal. At VC, however, a different picture emerges. Very few cells can be activated from the MD eye. In other words the ocular dominance (OD) histogram is skewed toward the nondeprived eye (see Figure 2, Section B). Binocularity, of course, is lost. This behavioral-electrophysiological constellation of visual deficiency is referred to as *amblyopia*. Once it is established, it is relatively permanent. Only slight behavioral and little or no electrophysiological amelioration is produced by subsequent binocular or forced monocular visual experience.

In contrast to the profound visual disturbances produced by even short periods of MD in kittens, prolonged MD in the adult cat is without significant effect. Wiesel and Hubel were the first to propose and demonstrate a developmental "critical period" during which manipulations of the environment could profoundly affect neural development, for example, 4–16 weeks of age from birth for the kitten visual system. Environmental manipulations outside of this period produce little or no effect. More subtle restrictions of the visual environment also produce modifications of visual system development, but only during comparable critical periods.

Two crucial experiments have extended our understanding of this process. Kittens, at birth, show visual systems possessing the basic elements of the normal adult groundplan. Most VC cells are binocular and crude, but basic elements of hierarchical RF organization are present. If kittens are binocularly deprived (BD) from birth by lid suture, visual abnormalities are found when testing takes place past 16 weeks. However, the result is not what one might expect as an extrapolation of MD findings. Although BD cats demonstrate major behavioral deficiences, they are not as profound as those seen in the deprived eye of MD kittens (Mower, Caplan, & Letsou, 1982; D. Smith, 1981). Moreover, one does not see a total absence of ability to drive VC neurons that one might predict from extrapolation of MD studies. Binocularity of input to single VC cells is preserved although diminished. Some elements of RF organization persist although individual RFs appear less well defined (Kratz, Spear, & Smith, 1976; Mower, Caplan, & Letsou, 1982; Wiesel & Hubel, 1965). As for MD, recovery from BD is incomplete (Cynader, Berman, & Hein, 1975; Kratz, Spear, & Smith, 1976).

From the standpoint of a single kitten's eye, it is better to be in a BD than in an MD paradigm. The presence of one eye receiving normal visual input (MD) produces graver consequences than if both eyes are deprived (BD). This fact has led to the hypothesis that both eyes compete for VC neurons. If balance is maintained through normal visual experience or through BD, then each eye ends up with equal influence. However, when an environmental ocular imbalance occurs, as produced by the MD paradigm, then the eye receiving the more normal input achieves the greater access to VC neurons. Neuroanatomical mapping studies strongly support this concept (Hubel, Wiesel, & Levay, 1977). The ocular dominance stripes determined on injection of the MD eye are greatly reduced in size, whereas those from the nondeprived eye are increased in size (see Figure 3, Sections B and C). There has been an apparent anatomical takeover of VC by the nondeprived eye (Levay, Wiesel, & Hubel, 1980; Shatz & Stryker, 1978).

Other restrictions of the visual environment have also been shown to produce anticipated modifications of the visual system, but only during comparable critical periods. Such experimental manipulations include wearing of optical prisms (E. L. Smith, Bennett, Harwerth, & Crawford, 1979), restricted visual environments that contain stripes of a single orientation or movement in a single direction (Blasdel, Mitchell, Muir, & Pettigrew, 1977; Creutzfeldt & Heggelund, 1975; Cynader et al., 1975; Daw & Wyatt, 1976; Stryker, Sherk, Leventhal, & Hirsch, 1978; Van Sluyters & Blakemore, 1973), and alternating MD and surgical strabismus (Hubel & Wiesel, 1965b; Ikeda & Tremain, 1977; Ikeda & Wright, 1976; Kratz & Spear, 1976).

2.1.3. Monocular Rotating Prism Amblyopia

Certain studies have been criticized because they require profound alteration of the environment to produce effects, for example, complete MD, viewing only one stripe orientation, cutting an extra ocular muscle, and so forth. The implications are that the experimental paradigms would not be relevant to human development where such extremes of environmental abnormality are not ordinarily experienced. To address this issue directly, we performed a series of experiments designed to determine whether the profound effects of MD could be reproduced without total unilateral pattern deprivation (Mower, Burchfiel, & Duffy, 1982). To investigate the necessary condition(s) for competitive takeover in VC, we modified the Hubel and Wiesel MD paradigm. Kittens were fitted with goggles containing lenses, One eye perceived the world in a normal fashion via plain glass, but the other eye saw the world through a rotatable wedge prism. Every 5 to 10 minutes during a daily two-hour period of binocular visual experience, the wedge was randomly rotated to a new position. The net result was that one eye experienced completely normal visual input. The manipulated eye experienced a pattern-rich retinal image, but with an unreliable relationship to external space resulting from the random prism rotations. We hoped the kittens would attend to the normal image and internally suppress the unreliable image from the manipulated eye just as the human child with strabismus learns to suppress the image from his unreliable "lazy eye." To our surprise electrophysiological investigations of monocular rotating prism (MRP) cats past 16 weeks revealed exactly that, namely, a takeover of VC cells by the normal eye comparable to that seen after MD (see Figure 2, Section C). Moreover, the cats appeared behaviorally deficient when the MRP eye was monocularly evaluated. Thus, in the presence of light and of pattern, an asymmetry of useful input clearly biased the competition for VC neurons in the direction of the eye receiving reliable visual information. Presumably, internal monocular suppresion functioned similarly to external MD to produce the shift of the OD curve. Such subtle influences on visual performance are well within what infants and young children may experience, for example, strabismus. Thus, at least for the visual systems, environmental manipulations on the order of what the infant may naturally experiencecan produce profound and presumably long lasting neurobehavioral changes.

2.1.4. Deprivation Free Environmental Effects

Most experimental demonstrations of visual system plasticity have employed one or another form of environmental restriction or limitation. Spinelli and Jensen (1979), however, took the opposite approach. They reported that early experience could produce plastic neural changes in the visual and somatic cortices of animals that were raised with periodic sensory overstimulation rather than conditions of constant sensory deprivation.

To demonstrate this effect, Spinelli and Jensen (1979) applied paired somatic and visual stimuli to kittens during the first few months of life. Each kitten received an electrical stimulus to one dorsal forearm that was paired with a visual pattern displayed through goggles: one pattern (vertical lines) was delivered to one eye with the shock, and a different pattern (horizontal lines) was presented to the other eye when there was no shock. The kitten could control the stimuli by lifting its forearm that deactivated the shock via a lever. Thus, the kittens were removed from a normal cat colony and trained for a few minutes a day with a task that was behaviorally relevant: the task required visuomotor coordination to avoid a mildly unpleasant sensory stimulus. The experiment attempted to superimpose a consistent daily stimulus complex analagous to the kind of environmentally relevant experiences that a developing animal might encounter.

Neurophysiological changes seen after this training were dramatic: the cortical topographic map of the receptive field for the stimulated area of the kittens' forearms was enlarged two to threefold in contrast to the untrained (control) forearm. In addition, RFs for the visual pattern orientations used during training strongly predominated over others in the visual and somatosensory cortices of these animals. A most unusual finding was the presence of VC cells demonstrating RF properties not seen in the normal, untrained cat. Many single VC cells that received binocular input demonstrated a different RF in each eye; one characteristically was optimized for response to vertical lines and the other to horizontal lines.

Thus, an additional experience, even delivered on a background of normal environmental stimulation, still had a profound effect on the allocation of cortical resources at an electrophysiological level. A follow-up experiment (Spinelli, Jensen, & DiPrisco, 1980) analyzed the anatomical characteristics of the areas of somatosensory cortex that had demonstrated the enlarged topographic map of the stimulated forearm. Histologic staining revealed increased dendritic branching in this area of cortex. Although these experiments were based on a limited training experience administered to kittens, their impact rests on the marked changes that even a simple experience appears to induce on cortical structure and function. The fact that the changes reported were not induced by deprivation but rather by active experience indicates that the allocation of cortical resources may correlate with the behavioral relevance of the various stimulus types present in the environment, especially during development.

If cortical receptivity and architecture can be selectively guided by the organism's experiences, and if this can be proven to have behavioral functional expression, then this finding has strong clinical implications not only for the developmental capabilities of the normal young, but also for the degree of success that can be expected from rehabilitation training after injury. These experiments have preliminarily opened the investigation of the significance of neuronal plasticity in the normal brain, without introducing the confounding effects of sensory deprivation.

2.1.5. Environmental Role in Commencement of Critical Period

Another question of considerable importance relates to the timing of the critical period itself. For the kitten, does it always begin shortly after eye opening (week 3) and gradually end by week 16? Is it genetically preprogrammed or is it triggered by some phenomenon such as first exposure to light? To investigate this possibility, we began a series of experiments designed to determine whether we could delay the onset of the critical period by preventing visual experience (Mower, Berry, Burchfiel, & Duffy, 1981). To begin with, binocular lid suture (BS) was performed on kittens near the time of spontaneous eye opening. At the end of the classic critical period (16 weeks), one eyelid was opened and the animal was given several months of MD experience. The thought was that, if the critical period had already ended, the animals would show the BS (or BD) neurophysiological pattern; however, if the critical period had been postponed, the MD paradigm should have a significant effect. The BS to MD preparation did not produce the classic behavioral or physiological effects of MD amblyopia. Animals were "blind" bilaterally and showed the neurophysiological pattern associated with BD. However, another set of kittens was reared in total darkness (DR) until 16 weeks and then brought into a normal environment after having one lid sutured shut. These two groups (BS and DR) permitted a comparison between the effects of diffuse light (BS) and complete light deprivation (DR), since the sutured lids do not eliminate light stimulation but merely attenuate it by 1–2 log units (Loop & Sherman, 1977). After a period of MD experience this DR to MD preparation did reproduce the classic cortical effects of MD despite the fact that the environmental manipulation occurred beyond the end of the classic critical period. Now, form vision is prevented from reaching the retina in both the BS and DR states. However, no light at all reaches the DR retina, whereas significant light reaches the retina of the BS kitten. The mere presence of light through the closed lids appeared sufficient to trigger the critical period in the BS preparation. In the DR kittens, the critical period began only at the point of first exposure to light.

Thus, the timing of the critical period is not uniquely under genetic control; it requires external environmental stimulation for initiation. If it can be postponed by prolonging complete darkness, can it be initiated early by premature exposure to light? These data may have direct implications for the preterm human infant. Is it possible that premature exposure to complex visual, auditory, and somesthetic stimuli triggers critical periods before they would start in the usual course of events. Perhaps the developmentally immature preterm infant cannot deal with his complex, stimulus rich environment during this sensitive period with the same security and facility he might otherwise have at the full-term point. Because of his relative physiological immaturity he may be unable to interact with incoming signals in the same complex manner as would be possible by term. Thus, for at least a portion of such a sensitive period, he may be disadvantaged, even deprived, of developmental interactions critical to normal development. Behavioral evidence suggests that premature infants actively withdraw from, or attempt to shut out, stimuli that are sociotaxic for full-term infants. In a sense, premature infants may actually deprive themselves during this sensitive period much like the MRP cats who shut out the unreliable, conflicting signal from one eye. Thus, a necessary protective reflex shutting out excessive stimulation to maintain physiological homeostasis may have the secondary consequence of sensorial and interactive deprivation. Could these postulated forms of deprivation be a contributory factor to the sensory and motor problems many premature infants manifest as they grow older?

2.2. Pharmacological Modification of Environmental Effects

2.2.1. Pharmacological Restoration of Function in MD Amblyopia: Neural Atrophy versus GABA Mediated Inhibition

The usual explanation for the functional disconnection between the MD eye and the VC has been atrophy of neural connections induced by selective disuse. Anatomical studies have, indeed, supported this view (Levay *et al.*, 1980; Shatz & Stryker, 1978). However, an alternative hypothesis is that the connections remain intact but are partially or completely *inhibited* from useful function. This would mean that the brain is not necessarily immutable beyond the critical period. In our laboratory we demonstrated that the apparent disconnection between the deprived eye and visual cortex in MD kittens is partially attributable to excessive inhibition—not only some form of neural atrophy (Burchfiel & Duffy, 1981; Duffy, Snodgrass, Burchfiel, & Conway, 1976; Duffy, Burchfiel, & Snodgrass, 1978).

Clinical experience with amblyopic patients (children and adults) suggests that the visual deficit fluctuates and is not as fixed as the above-mentioned anatomical evidence might suggest. In 1921, Amann first noted that the wearing of dark glasses was less harmful to the acuity of the amblyopic eye than the normal eye. In the 1950s and 1960s, Von Noorden and others (Van Noorden, 1961; Von Noorden & Burian, 1959a,b, 1960; Von Noorden & Leffler, 1966) observed that at low levels of illumination, the acuity of the amblyopic eye is equal to, or occasionally better than, that of the same individual's normal eye. He suggested that the effects of low levels of illumination might be to reduce the levels of contrast. In agreement with this, it was shown that the amblyopic eye performs better when the background is not cluttered with extraneous pattern. Furthermore, in the presence of stimulation of the normal eye, the amblyopic eye showed a degradation of performance. The presence of such fluctuation in amblyopia did not appear consistent with a fixed anatomic model as had been suggested by the anatomical data. Some years ago, therefore, we undertook a study of monocular visual evoked response (VER) to plain and patterned light in children with established amblyopia (Lombroso, Duffy, & Robb, 1969). In the normal eye, we saw the expected increase in the VER wave-form amplitude and complexity with the addition of pattern to the stimulus. However, for stimulation of the amblyopic eye, the introduction of pattern produced a decrease in wave-form amplitude and complexity. It appeared that the amblyopic eye and its connected visual circuitry *were* capable of detecting pattern, but that when pattern was perceived, further visual processing was shut down, resulting in a paradoxic VER reduction. Thus, we postulated that pattern was capable of inducing inhibitory suppression of visual processing in amblyopia and that excessive suppression or inhibition might be a major physiologic defect in amblyopia.

To investigate this inhibitory hypothesis (Duffy et al., 1976; Duffy et al., 1978), we turned to the MD kitten model of amblyopia. The major inhibitory neurotransmitter in cat visual cortex appears to be gamma-aminobutyric acid (GABA) (Hunt, 1982; Pettigrew & Daniels, 1973; Sillito, 1975a). We hypothesized that the administration of bicuculline, a GABA antagonist, might restore cortical input from the MD eye if GANA-mediated inhibition were a major factor. Bicuculline would have little or no effect if atrophy were the sole factor. Kittens were deprived of vision at 3 weeks of age and so maintained for 9 months, well past the end of the critical period. When the deprived eye was opened and unilaterally stimulated at age 9 months, single-unit, extracellular, microelectrode studies revealed the same deficit as seen by Hubel and Wiesel, that is, no activation of cortical neurons by the deprived eye. We then administered intravenous bicuculline at levels insufficient to cause convulsions—a common side effect for larger doses. As we hypothesized, a large number of cells that could not be previously activated from the deprived eye could now be driven for varying periods of time (see Figure 5). It thus became possible to open the connection between the deprived eye and visual cortex to approximately 60% of all cells encountered. Furthermore, the RF characteristics of these cells appeared identical to those of the normal eye connected to that cell when individually tested. Antagonists of other putative inhibitory neurotransmitters, such as strychnine, and purely excitatory agents, such as glutamate, failed to produce functional reconnection between the MD eye and VC neurons. Thus, the effect was specific to antagonism of GABA. More recently, we applied bicuculline directly to VC cells using the technique of microiontophoresis (Burchfiel & Duffy, 1981) and demonstrated over a 40% restoration of input to VC from the MD eye of cats deprived for many months (see Figure 6). These findings were confirmed by Sillito, Kemp, and Blakemore (1981). Blakemore, Hawken, and Mark (1982) demonstrated, from intracellular recordings in VC of MD cats, that neurons receive excitatory input from the deprived eye with RF organization corresponding to that of the nondeprived eye even though this input is not sufficient to produce firing of the neuron. After a very brief period of MD, the iontophoretic application of excitatory agents to kitten VC may open inputs to the deprived eye (Blakemore & Hawken, 1982; Blakemore et al., 1982); however, after many months of deprivation, only disinhibitory agents, such as bicuculline, can restore the MD eye to VC connection (Burchfiel & Duffy, 1981; Sillito et al., 1981). Thus, the dramatic loss of cortical activation produced by MD cannot be completely explained by atrophy, but must result, at least in part, from excessive inhibition of otherwise intact circuitry. This concept is supported by work demonstrating that many VC

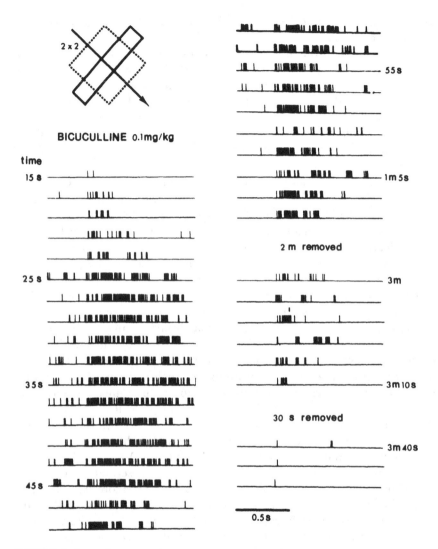

FIGURE 5. Bicuculline reversal of MD amblyopia. A demonstration of the ability of intra-venous bicuculline, an antagonist of the inhibitory neurotransmitter GABA, to restore functional connectivity between the deprived eye and the VC of an MD cat. At top left is illustrated the RF of a cat VC neuron recorded following 9 months of monocular deprivation. The cell responded best to a moving slit in the direction indicated by the arrow, when the normal eye was stimulated, however, the cell did not fire to any stimuli delivered to the deprived eye in the baseline state. Response of deprived eye for approximately four min. after 0.1 mg/kg intravenous bicuculline is shown. Each horizontal line is a response to one stimulation. Each small vertical line represents one cell firing. Note development of response just beginning by 15 sec., reaching peak by 40 sec., and disappearing by 3 min. 44 sec. Moreover, the RF in the deprived eye was identical to that of the nondeprived, normal eye.

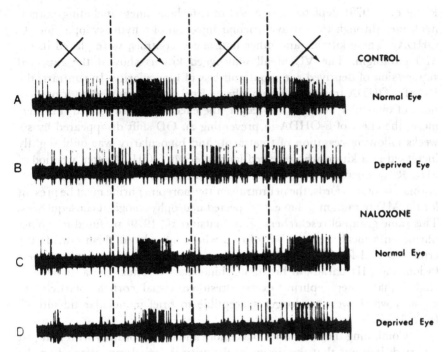

FIGURE 6. Direct, microiontophoretic application of bicuculline. Restoration of binocular input to a VC neuron of a cat who had been monocularly deprived for many months from birth. The cell had an RF in the nondeprived eye that preferred a vertical line stimulus moving to the right. Movement to the left or of a horizontal line up and down produced no response (A). Stimulation of the deprived eye produced no response (B). Five minutes after the start of bicuculline application, the cell began to fire from deprived eye stimulation (D). The unidirectional response to movement of the optimum stimulus (A) was altered by bicuculline to a bidirectional response in both eyes (C and D). In this case, restoration of the connection from the MD eye to VC was obtained with a slight degredation of RF stimulus specificity. This study (Burchfiel & Duffy, 1981) and others (Sillito *et al.*, 1981) establish the locus of amblyopia to be the VC and emphasize the role of inhibition in the maintenance of the monocular visual dysfunction that characterizes amblyopia. From *Brain Research*, 1981, *206*, 479–484. Copyright 1981 by Elsevier Biomedical Press B. V., Amsterdam, the Netherlands. By permission of the author and the publisher.

cells can be driven by the deprived eye of mature MD cats following enucleation of the normal eye (Kratz *et al.*, 1976).

2.2.2. *Pharmacological Prevention of MD Amblyopia: Norepinephrine Control of Plasticity*

Recent work by others has implicated norepinephrine as yet another neurotransmitter having an impact on brain plasticity. Kasamatsu and

Pettigrew (1979) depleted kitten VC of catecholamines (including norepinephrine) through the intraventricular injection of 6-hydroxydopamine (6-OHDA). These kittens, and other undrugged controls, were placed in the MD paradigm. The VC of all undrugged kittens showed the expected suppression of deprived eye input and loss of binocularity. In contrast, the VC of 6-OHDA treated (norepinephrine free) kittens showed high proportions of binocular neurons. Moreover, the effect was dose related. Furthermore, the effect of 6-OHDA in preventing an OD shift disappeared by six weeks following cessation of treatment, and binocularity was only slightly increased in a kitten who received large doses of 6-OHDA after a period of MD. RF properties of binocular cells in 6-OHDA treated MD cats were normal. In other words, the neurotransmitter norepinephrine must be present for the MD paradigm to have its expected neurophysiological consequences. This same group of researchers (Kasamatsu et al., 1979) perfused norepinephrine onto local areas in VC of cats whose age placed them outside the critical period for MD. Nonetheless, in areas of local norepinephrine perfusion, the MD paradigm broke VC binocularity. Thus, the authors concluded that norepinephrine excess "restored visual cortical plasticity in animals which were no longer susceptible to brief monocular lid-suture" (p. 163).

Combining the work from Blakemore and our laboratory with that just reported, it seems that the timing of the critical period may depend on the neurotransmitter norepinephrine, and the maintenance of the consequences of experience during the critical period may depend on another neurotransmitter, namely, GABA.

2.2.3. Possible Effects of Endogenous Opioids on Sensory Neural Development

Although all evidence is not yet firmly established, it appears that organisms respond to stress at least in part via the release of endogenous opioids (Herman & Panksepp, 1981; J. W. Lewis, Cannon, & Liebeskind, 1980; MacLennan, Drugan, Hyson, Maier, Madden, & Barchas, 1982; Madden, Akil, Patrick, & Barchas, 1977; Parker, Simpson, Bilheimer, Leneno, Carr, & MacDonald, 1980). This effect may be enhanced by other factors such as corticosterone (MacLennan et al., 1982). Cerebral cortex contains opiate receptors (Hiller, Pearson, & Simon, 1973; Kuhar, Pert, & Snyder, 1973) and opiate peptides (Simantov, Kuhar, Pasternak, & Snyder, 1976; Yang, Hong, & Costa, 1977) and responds electrically to opiate peptides (Nicoll, Siggins, Ling, Blom, & Guillemin, 1977; Palmer, Morris, Taylor, Stewart, & Hoffer, 1978). It seems reasonable to presume, therefore, as Als and Duffy have suggested earlier (1983b) that premature infants would employ this intrinsic mechanism to obviate the distress produced by environmental overstimulation. Indeed, mother's milk contains morphinelike material (Hazum, Sabatka, Chang, Brent, Findlay, & Cuatrecasas, 1981) and may constitute an extrinsic source of stress relief beyond that of hunger. Als and Duffy (1983b) have furthermore speculated about the evidence that now suggests that morphinelike substances may be involved in sensory integrative processes when considering the premature human organism. Neurons of rat frontal cortex carry multiple opiate receptors (Williams & Zieglgansberger, 1981). In monkey cerebral cortex, morphine-type receptors increase in quantity in a gradient along hierarchically organized cortical systems that process sensory information of a progressively more complex nature (M. E. Lewis, Mishkin, Bragin, Brown, Pert, & Pert, 1981). These authors speculate that "opiate receptors may play a role in the affective filtering of sensory stimuli at the cortical level, that is, emotion-induced selective attention" (p. 1166). Wise and Herkenham (1982) similarly suggested that opiates might be important in the modulation of specific cortical elements. They speculated that "opiates may predominantly influence the outflow of cortical fields and those fields involved in polymodal information processing and limbic functions" (p. 389).

Therefore, if stress in the premature infant produces parallel changes in central opioid levels as it does in the rat (Madden, Akil, Patrick, & Barchas, 1977), then an alteration of cortical function and development might be expected. Direct evidence comes from the work of Vorhees (1981) who exposed pregnant rats to the morphine antagonist naloxone and determined the consequence on their offspring. Naloxone-exposed progeny were accelerated in early growth, righting development, startle development, home-scent discrimination, and in directional swimming development. However, as adults they showed impaired maze learning.

In our own work, we found that naloxone, like bicuculline, was capable of restoring input from the deprived eye of MD cats to VC (Duffy et al., 1978; see also Figure 7). Moreover, it has been shown to functionally improve deprived eye vision in a behavioral task (Mower, Caplan, Burchfiel, & Duffy, 1982). The naloxone effects differ from those of bicuculline by taking much longer to become evident, requiring high dose schedules; yet also lasting much longer.

Although the role of the opiates is not yet clearly understood, it is reasonable to presume that excesses of either morphine agonists or antagonists in the critical periods of sensory development may have adverse consequences for future function. Thus, a mechanism appropriate for stress reduction at older age points may have adverse consequences if called into-play during early sensory-neural development.

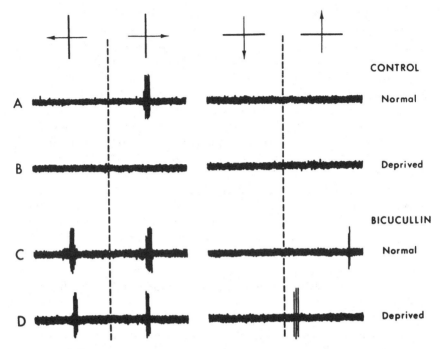

FIGURE 7. Reversal of MD effects by intravenous naloxone. Baseline response of a single cell in the VC of a cat, monocularly deprived of vision for many months from birth, to stimulation of a normal eye (A) and deprived eye (B). Following a large dose of naloxone, a response to stimulation of the deprived eye was clearly evident (D) at which time the response to stimulation of the normal eye was clearly maintained (C). The RF in both eyes preferred a linear stimulus moving down to the right at 45°. Upward movement or movement 90° to this produced no response before or after naloxone. In this case, reversal of MD was obtained with the loss of baseline RF properties. The pharamacological action by which naloxone restores MD eye input is not known. Naloxone is primarily a morphine antogonist but may have some anti-GABA properties. Despite the uncertain pharmacological mechanism, the restorative action of naloxone in experimental MD amblyopia is of interest for it occurs at levels that produce no untoward clinical side effects. Bicuculline, on the other hand, produces a restoration of MD eye input at drug levels not far from those that produce dangerous convulsions. Because naloxone is relatively safe, its behavioral effects are now being studied in performing animals. It appears to produce a significant restoration of functional acuity in the deprived eye (Mower, Caplan, Burchfiel, & Duffy, 1982).

3. FACTORS RESPONSIBLE FOR RECOVERY FROM EXPERIMENTAL LESIONS

So far we have emphasized the impact of environment on brain devel-opment in normal, unlesioned animals. However, there is a large body of literature dealing with restoration of central nervous system function and anatomy in animals following specific brain lesions (see, for example, the review by Cotman & Nieto-Sampedro, 1982). To some degree the environment appears to play a role in this recovery process.

3.1. Role of Training in Recovery from Lesions

As emphasized by Goldman (1976; Goldman & Lewis, 1978) return of function in rhesus monkeys lesioned in early life may depend heavily on experience. For example, orbital prefrontal lesions in adult monkeys cause severe impairments on spatial alternation tests. When the lesion is placed in infancy, however, performance may reach normal levels by two years of age. However, this recuperation requires prior training on a series of cognitive tasks during development (Goldman, 1976). Of interest is the fact that such recovery of spatial ability may be stimulated by training that actually reinforces nonspatial strategies (Goldman & Lewis, 1978). Thus, the beneficial effects of environmental stimulation may, in some circumstances, act nonspecifically. In other circumstances the environment may play a more limited role. Whereas rearing brain-injured rats in enriched environments improves maze performance (Schwartz, 1964; Will, Rosenzweig, Bennett, Herbert, & Morimoto, 1977), similar forced enrichment does not improve visual discrimination after total VC lesions (Bland & Cooper, 1969). However, Goldman and Lewis (1978) suggest that when the environment fails to foster recovery, lesions are generally extensive or involve critical regions.

3.2. Mechanisms of Lesion Recovery

Many mechanisms may be involved in the environmentally fostered recovery process. Normal rats reared in enriched environments have thicker and heavier cortices (Rosenzweig, Krech, Bennett, & Diamond, 1962), shorter latency, greater amplitude photic-evoked potentials (Edwards, Barry, & Wyspianski, 1969; Mailloux, Edwards, Barry, Rowsell, & Achorn, 1974; Myslivecek & Stipek, 1979), and enhanced metabolic activation, especially cellular RNA content (Ferchmin, Eterovic, & Caputto, 1970; Uphouse & Bonner, 1975). It is generally believed that elucidation of the mechanisms underlying this anatomical, electrical, and biochemical facilitation in normal animals will explain environmentally fostered recovery in lesioned animals.

Stimulation produces enhanced production of dendritic branches and spines and greater synaptic number and/or efficiency (Berard, Burgess, & Coss, 1981; Diamond, Krech, & Rosenzweig, 1964; Kawaguchi, Yamamoto, Samejima, Itoh, & Mizuno, 1979; Por, Bennett, & Bondy, 1982; Vrensen & Cardozo, 1981; Wesa, Chang, Greenough, & West, 1982). It is believed that such mechanisms are responsible for reinnervation of neurons denervated by nearby lesions and for increased sensitivity of partially denervated neurons. Goldman and Lewis (1978) nicely summarize this process by suggesting that "rich or varied experience during development stimulates trophic responses in brain tissue, and further, that these trophic responses form the basis of the physiological events underlying behavioral recovery from early (and perhaps later) brain injury" (p. 300).

The specific, restorative nature of this process is no better illustrated than by the work of Merzenich and Kaas (1982) who demonstrated the phenomenal ability of the primate somatosensory cortex to reorganize following peripheral nerve or spinal cord lesions. Following finger denervation, the cortical representations of the remaining enervated fingers enlarge especially those areas most frequently used by the animal.

It is now known that excitation of the postsynaptic membrane by stimulation-released neurotransmitter has an action beyond that of simple transfer of information (spikes) from one neuron to another. Through a complex series of biochemical reactions triggered by the neurotransmitter and involving cyclic AMP or GMP, an increased rate of synthesis of ribosomal RNA results and, subsequently, an increase of building-block proteins (Goldman & Lewis, 1978) in the postsynaptic neuron.

However, substances other than neurotransmitters are believed to exert trophic influences on neuronal tissue. One interesting line of research suggests that in both the normal (Crutcher & Collins, 1982) and lesioned (Nieto-Sampedro, Lewis, Cotman, Manthorpe, Skaper, Barbin, Longo, & Varon, 1982) brain, "growth" factors may be isolated. Small pieces of Gelfoam within lesions placed in developing rat brains yield substances with marked neuronotrophic activity when assayed on tissue-cultured neurons (Nieto-Sampedro *et al.*, 1982).

Such trophic influences may serve to facilitate the establishment of synapses between appropriate axonal and dendritic partners. According to Purves and Lichtman (1980), synapse establishment depends on mechanisms of cellular differentiation, cellular migration, and axonal guidance. However, an important question that has recently come under investigation remains: How does the brain regulate synaptic connectivity (Crepel, 1982; Purves and Lichtman, 1980)? For instance, in the primate each motor axon contacts more muscle fibers at birth than at maturity. It is now believed that elimination of some initial connectivity in the postnatal period is not limited to muscle but occurs within the brain as well. The basic process is not necessarily a reduction of synapses, rather a reduction of the number of different axons innervating each cell. Synaptic numbers, of course, increase after birth. The net consequence is the progressive confinement of the endings on a particular axon to a smaller number of target cells. No doubt competition between competing axons is motivated by trophic factor. As one axon terminal gains access to the target cell, it enhances its share of trophic factor at the expense of its competitors. To some degree, neural activity plays a role, for it has been shown that synchronous electrical activity in competing axons impedes, whereas asynchronous activity facilitates synapse elimination. This progressive confinement of axonal-neuronal contact may well explain the decreased malleability of the brain with age. From the other perspective, the enhanced, redundant, synaptic congestivity seen at birth is without question partially (at least) responsible for the greater ability to compensate for lesions in the neonatal period than at later ages.

3.3. Prenatal Lesions in Monkeys

Recent advances in surgical techniques have permitted lesions to be placed in fetal monkeys with the anatomical and functional consequences determined following normal birth at gestational term. According to Goldman-Rakic (1980a,b) fetal lesions do not notably prolong or reinitiate neuronal proliferation after these processes normally cease. It appears that fetal lesions, especially of the frontal lobe, involve a change in the migratory course of neurons to populate regions adjacent to the lesioned area. Clearly, this compensatory mechanism would not be available to fetal monkeys beyond the stage of completion of cortical migration. Influence of the fetal environment on either normal or on compensatory fetal cell migration is not known (Rakic, 1983). However, this paradigm promises to teach us much about plasticity of the fetal primate brain.

4. SUMMARY

As we have seen, there is much evidence from the animal laboratory to support the notion that brain development is a mutual process, responding to alterations of both internal and external milieu. Environmental enrichment in developing animals can increase brain size, complexity, and function. Even nonspecific training may hasten recovery from certain specific experimental lesions. These findings support the hopes of the special educator or the physical therapist. Whereas it is generally agreed that animal experiments demonstrate a definite and, in many ways, surprisingly specific interaction between the environment and brain development, they have been used to bolster opposing viewpoints. One school of researchers sees the animal work as an affirmation of the profound effect that the environment has on the development of brain and behavior. They emphasize the potential benefits of supportive measures and intervention. Another perspective, in contrast, emphasizes the degree of environmental distortion necessary to produce an effect. It emphasizes the immutable nature of brain and behavior outside the critical period and the difficulty in defining the bounds of such a period in infants. These researchers see human behavior and infant development as largely genetically determined, requiring only the absence of extreme environmental deviation to achieve predestined potential. Both schools would agree that it is a big step from experimental animal work to therapeutic work with humans, and that there is much to learn.

Animal experimentation is, to a large extent, geared toward improving our understanding of natural recuperative mechanisms so as to suggest new clinical strategies for our work with humans. For example, pharmacological release of inhibition, such as that which transiently restored vision to the amblyopic eye of MD cats, may eventually prove useful in other neurological conditions. "Fixed" neurological deficits need not necessarily imply absent compensatory mechanisms. Alternate neural circuits may simply be strongly inhibited. Training under disinhibitory drugs may allow initiation of recuperative processes otherwise inaccessable by conventional means. Much work lies ahead to elucidate the still missing steps projected from the animal data.

But the animal literature also brings warnings, especially for our management of the human premature infant. It appears possible that the impact of polysensory stimulation prior to term may trigger a sensitive period at a time when the organism cannot appropriately incorporate such sensory influences. Moreover, strenuous medical management and/or programs of environmental enrichment may force the organism to reject such stimulation with consequences akin to the MRP kitten. If stimulation proves too stressful, internal relief via release of endogenous opioids might ensue, which could have serious consequences for neural development. Does this mean that a premie's environment should mimic the uterus—dark, swaddled, and quiet? Or is there no way to avoid triggering a sensitive period of reactivity in premies? Would return to a darkened environment, therefore, be deleterious? No doubt the optimal environment should fall somewhere between total sensory deprivation and continual forced and often contingent enrichment. Only careful environmental manipulation in controlled studies with sensitive outcome measures will provide the answers. What the animal literature tells us is that the premature infant's environment may well have a profound effect on subsequent neural development. The premie's milieu needs careful study and should not consist of a random sequence of medical events as it so often does.

5. REFERENCES

Als, H. Towards a synactive theory of development: Promise for the assessment and support of infant individuality. *Infant Mental Health Journal*, 1983, *3*, 229–243.

Als, H., & Duffy, F. H. Assessment of premature infants behavior. In T. B. Brazelton & B. M. Lester (Eds.), *New approaches to developmental screening of infants*. New York: Elsevier, 1983. (a)

Als, H., & Duffy, F. H. Conceptualizing and assessing the behavior of the fetal newborn. In D. S. Staton (Ed.), *Caring for special babies*. Chapel Hill: University of North Carolina Press, 1983, in press. (b)

Als, H., Lester, B. M., Tronick, E., & Brazelton, T. B. Toward a research instrument for the assessment of preterm infants' behavior (A.P.I.B.). In H. E. Fitzgerald, B. M. Lester, & M. W. Yogman (Eds.), *Theory and research in behavioral pediatrics* (Vol. 1). New York: Plenum Press, 1982. (a)

Als, H., Lester, B. M., Tronick, E., & Brazelton, T. B. Manual for the assessment of preterm infants' behavior (A.P.I.B.). In H. E. Fitzgerald, B. M. Lester, & M. W. Yogman (Eds.), *Theory and research in behavioral pediatrics* (Vol. 1). New York: Plenum Press, 1982. (b)

Amann, E. Eurige Beobachtungen bei den Funktionsprufungen in der Sprechstunde. "Zentrales" Schen-Schender Glaukomatosen- Schen der Amblyopen. *Klinische Monatsblaetter fr Augenkunde*, 1921, *67*, 564–573.

Amiel-Tison, C., Barrier, G., Shnider, S., Levinson, G., Hughes, S. C., & Stefani, S. J. The neonatal neurologic and adaptive capacity score (NACS). *Anesthesiology*, 1982, *56*, 492–493.

Baker, M. C., Spitler-Nabors, K. J., & Bradley, D. C. Early experience determines song dialect responsiveness of female sparrows. *Science*, 1981, *214*(13), 819–821.

Barlow, H. B. David H. Hubel and Torsten Wiesel: Their contributions towards understanding the primary visual cortex. *Trends in Neurosciences*, 1982, *5*, 145–152.

Berard, D. R., Burgess, J. W., & Coss, G. Plasticity of dendritic spine formation: A state-dependent stochastic process. *International Journal of Neuroscience*, 1981, *13*, 93–98.

Blakemore, C., & Hawken, M. J. Rapid restoration of functional input to the visual cortex of the cat after brief monocular deprivation. *Journal of Physiology* (London), 1982, *327*, 463–487.

Blakemore, C., Hawken, M. J., & Mark, R. F. Brief monocular deprivation leaves subthreshold synaptic input on neurons of cat's visual cortex. *Journal of Physiology* (London), 1982, *327*, 489–505.

Bland, B. H., & Cooper, R. M. Posterior neodecortication in the rat: Age at operation and experience. *Journal of Comparative and Physiological Psychology*, 1969, *69*, 345–354.

Blasdel, G. G., Mitchell, D. E., Muir, D. W., & Pettigrew, J. D. A physiological and behavioral study in cats of the effect of early visual experience with contours of a single orientation. *Journal of Physiology*, 1977, *265*, 615–636.

Brazelton, T. B. *Neonatal behavioral assessment scale*. Philadelphia: Lippincott, 1973.

Burchfiel, J. L., & Duffy, F. H. Role of intracortical inhibition in deprivation amblyopia reversal by microiontophoretic bicuculline. *Brain Research*, 1981, *206*, 479–484.

Cotman, C. W., & Nieto-Sampedro, M. Brain function, synapse renewal, and plasticity. *Annual Review Psychology*, 1982, *33*, 371–401.

Crepel, F. Regression of functional synapses in the immature mammalian cerebellum. *Trends in Neurosciences*, 1982, *5*, 266–269.

Creutzfeldt, O. D., & Heggelund, P. Neural plasticity in visual cortex of adult cats after exposure to visual patterns. *Science*, 1975, *188*, 1025–1027.

Crutcher, K. A., & Collins, F. In vitro evidence for two distinct hippocampal growth factors: Basis of neuronal plasticity? *Science*, 1982, *217*, 67–68.

Cynader, M., Berman, N., & Hein, A. Recovery of function in cat visual cortex following prolonged deprivation. *Experimental Brain Res.*, 1976, *25*, 139–156.

Daum, C., Grellong, B., Kurtzberg, D., & Vaughn, H. G. *The Albert Einstein neonatal neurobehavioral scale manual*. Unpublished manuscript, 1977. (Available from C. Daum, Department of Pediatrics, Albert Einstein College of Medicine, Bronx, New York, 10461.)

Daw, N. W., & Wyatt, H. J. Kittens reared in a unidirectional environment—evidence for a critical period. *Journal of Physiology* (London), 1976, *257*(1), 155–170.

Diamond, M. C., Krech, D., & Rosenzweig, M. R. The effects of an enriched environment on the histology of the rat cerebral cortex. *Journal of Comparative Neurology*, 1964, *123*, 111–120.

Drillien, C. M., Thomson, A. J. M., & Bargoyne, K. Low birthweight children at early school-age: A longitudinal study. *Developmental Medicine and Child Neurology*, 1980, *22*, 26–47.

Dubowitz, L., & Dubowitz, V. *The neurological assessment of the preterm and full-term newborn infant.* Clinics in Developmental Medicine, Spastics International Medical Publication (No. 79). Philadelphia: Lippincott, 1981.

Duffy, F. H. Topographic display of evoked potentials: Clinical applications of brain electrical activity mapping (BEAM). *Annals of New York Academy of Science*, 1982, *388*, 183–196.

Duffy, F. H., & Als, H. Neurophysiological assessment of the neonate: An approach combining brain electrical activity mapping (BEAM) with behavioral assessmant (APIB). In T. B. Brazelton & B. M. Lester (Eds.), *New approaches to developmental screening of infants.* New York: Elsevier, 1983.

Duffy, F. H., Burchfiel, J. L., & Lombroso, C. T. Brain electrical activity mapping (BEAM): A method for extending the clinical utility of EEG and evoked potential data. *Annals of Neurology*, 1979, *5*, 309–321.

Duffy, F. H., Burchfiel, J. L., & Snodgrass, S. R. The pharmacology of amblyopia. *Archives of Ophthalmology*, 1978, *85*, 489–495.

Duffy, F. H., Denckla, M. B., Bartels, P. H., & Sandini, G. Dyslexia: Regional differences in brain electrical activity by topographic mapping. *Annals of Neurology*, 1980, *7*(5), 412–419.

Duffy, F. H., Snodgrass, R., Burchfiel, J. L., & Conway, J. L. Bicuculline reversal of deprivation amblyopia in the cat. *Nature* (London), 1976, *260*, 256–257.

Edwards, H., Barry, W., & Wyspianski, J. Effects of differential rearing on photic evoked potentials and brightness discrimination in the albino rat. *Developmental Psychobiology*, 1969, *2*, 133–138.

Ferchmin, P. A., Eterovic, V. A., & Caputto, R. Studies of brain weight and RNA content after short periods of exposure to environmental complexity. *Brain Research*, 1970, *20*, 49–57.

Goldman, P. S. The role of experience in recovery of function following orbital prefrontal lesions in infant monkeys. *Neuropsychologia*, 1976, *14*, 401–412.

Goldman, P. S., & Lewis, M. E. Developmental biology of brain damage and experience. In C. Cotman (Ed.), *Neuronal plasticity.* New York: Raven Press, 1978.

Goldman-Rakic, P. S. Morphological consequences of prenatal injury to the primate brain. *Progress in Brain Research*, 1980, *53*, 3–19. (a)

Goldman, Rakic, P. S. Development and plasticity of primate frontal association cortex. In F. O. Schmitt & F. G. Worden (Eds.), *The cerebral cortex.* Cambridge: M.I.T. Press, 1980. (b)

Hazum, E., Sabatka, J. J., Chang, K. J., Brent, D. A., Findlay, J. W.A., Cuatrecasas, P. Morphine in cow and human milk: Could dietary morphine constitute a ligand for specific morphine receptors? *Science*, 1981, *213*, 1010–1012.

Herman, B. H., & Panksepp, J. Ascending endorphin inhibition of distress vocalization. *Science*, 1981, *211*, 1060–1062.

Hiller, J. M. & Simon, E. J. Inhibition by levorphanol of the induction of acetylcholinesterase in a mouse neuroblastoma cell line. *Journal of Neurochemistry*, 20 1973, 1789–1792.

Hubel, D. H., & Freeman, D. C. Projection into the visual field of ocular dominance columns in macaque monkey. *Brain Research*, 1977, *122*, 336–343.

Hubel, D. H., & Wiesel, T. N. Receptive fields of single neurons in the cat's striate cortex. *Journal of Physiology*, 1959, *148*, 574–591.

Hubel, D. H., & Wiesel, T. N. Single unit activity in lateral geniculate body and optic tract of unrestrained cats. *Journal of Physiology*, 1960, *150*, 91–104. (a)

Hubel, D. H., & Wiesel, T. N. Receptive fields of optic nerve fibres in the spider monkey. *Journal of Physiology*, 1960, *154*, 572–580. (b)

Hubel, D. H., & Wiesel, T. N. Receptive fields, binocular interaction, and functional architecture in the cat's visual cortex. *Journal of Physiology*, 1962, *160*, 106–154.

Hubel, D. H., & Wiesel, T. N. Shape and arrangement of columns in the cat's striate cortex. *Journal of Physiology*, 1963, *165*, 559–568.

Hubel, D. H., & Wiesel, T. N. Receptive fields and functional architecture in two nonstriate visual areas (18 and 19) of the cat. *Journal of Neurophysiology*, 1965, *28*, 229–289. (a)

Hubel, D. H., & Wiesel, T. N. Binocular interaction in striate cortex of kittens reared with artificial squint. *Journal of Neurophysiology*, 1965, *28*, 1041–1059. (b)

Hubel, D. H., & Wiesel, T. N. Receptive fields and functional architecture of monkey striate cortex. *Journal of Physiology*, 1968, *195*, 215–243.

Hubel, D. H., Wiesel, T. N., & Levay, S. *Plasticity of ocular dominance columns in the monkey striate cortex. Philosophical Transactions of the Royal Society of London*, 1977, B278, 377–409.

Hunt, S. P. GABA produces excitement in the visual cortex. *Trends in Neurosciences*, 1982, *5*, 101–102.

Ikeda, H., & Tremain, K. E. Different causes for amblyopia and loss of binocularity in squinting kittens. *Journal of Physiology* (London), 1977, *269*(1), 26–27.

Ikeda, H., & Wright, M. J. Properties of LGN cells in kittens reared with convergent squint: A neurophysiological demonstration of amblyopia. *Experimental Brain Research*, 1976, *25*, 63–77.

Kasamatsu, T., & Pettigrew, J. D. Preservation of binocularity after monocular deprivation in the striate cortex of kittens treated with 6-hydroxydopamine. *Journal of Comparative Neurology*, 1979, *185*, 139–162.

Kasamatsu, T., Pettigrew, J. D., & Ary, M. Restoration of visual cortical plasticity by local neuroprofusion of norepinephrine. *Journal of Comparative Neurology*, 1979, *185*, 163–182.

Kawaguchi, S., Yamamoto, T., Samejima, A., Itoh, K., & Mizuno, N. Morphological evidence for axonal sprouting of cerebellothalamic neurons in kittens after neonatal hemicerebellectomy. *Experimental Brain Research*, 1979, *35*, 511–518.

Kratz, K. E., & Spear, P. D. Effects of visual deprivation and alternations on binocular competition on responses of striate cortex neurons in the cat. *Journal of Comparative Neurology*, 1976, *170*, 141–152.

Kratz, K. E., Spear, P. D., & Smith, D. C. Post-critical period reversal of effects of monocular deprivation on striate cortex cells in the cat. *Journal of Neurophysiology*, 1976, *39*, 501–511.

Kuhar, M. J., Pert, C. B., & Synder, S. H. Regional distribution of opiate receptor binding in monkey and human brain. *Nature* (London), 1973, 245–447.

Levay, S., Wiesel, T. N., & Hubel, D. H. The development of ocular dominance columns in normal and visually deprived monkeys. *Journal of Comparative Neurology*, 1980, *191*, 1–5.

Lewis, J. W., Cannon J. T., & Liebeskind, J. C. Opioid and nonopioid mechanisms of stress analgesia. *Science*, 1980, *208*, 623–625.

Lewis, M. E., Mishkin, M., Bragin, E., Brown, R. M., Pert, C. B., & Pert, A. Opiate receptor gradients in monkey cerebral cortex: Correspondence with sensory processing hierarchies. *Science*, 1981, *211*, 1166–1169.

Lombroso, C. T. Neurophysiological observations in diseased newborns. *Biological Psychiatry*, 1975, *10*, 527–539.

Lombroso, C. T. Quantified electrographic scales on 10 preterm healthy newborns followed up to 40–43 weeks of conceptual age by serial polygraphic recordings. *Electroencephalography and Clinical Neurophysiology* (Amsterdam), 1979, *46*, 460–474.

Lombroso, C. T., Duffy, F. H., & Robb, R. M. Selective suppression of cerebral evoked potentials to patterned light in amblyopia exanopsia. *Electroencephalography and Clinical Neurophysiology* (Amsterdam), 1969, *27*, 238–247.

Loop, M. S., & Sherman, S. M. Visual discriminations during eyelid closure in the cat. *Brain Research*, 1977, *128*, 328–339.

MacLennan, A. J., Drugan, R. C., Hyson, R. L., Maier, S. F., Madden, J., & Barchas, J. D. Corticosterone: A critical factor in an opiod form of stress-induced analgesia. *Science* 1982, *215*, 1530–1532.

Madden IV, J., Akil, H., Patrick, R. L., & Barchas, J. D. Stress-induced parallel changes in central opioid levels and pain responsiveness in the rat. *Nature*, 1977, *265*, 358–360.

Mailloux, J. G., Edwards, H. P., Barry, W. F., Roswell, H. C., & Achorn, E. G. Effects of differential rearing on cortical evoked potentials of the albino Rat. *Journal of Comparative and Physiological Psychology*, 1974, *87*, 475–480.

Marler, P., & Peters, S. Sparrows learn adult song and more from memory. *Science*, 1981, *213*, 780–782.

Marx, J. L. How the brain controls birdsong. *Science*, 1982, *217*, 1125–1126.

Merzenich, M. M., & Kaas, J. H. Reorganization of mammalian somatosensory cortex following peripheral nerve injury. *Trends in Neurosciences*, 1982, *5*, 434–436.

Mower, G. D., Caplan, C., & Letsou, G. Behavioral recovery from binocular deprivation in the cat. *Behavioral Brain Research*, 1982, *4*, 209–215.

Mower, G. D., Berry, D., Burchfiel, J. L., & Duffy, F. H. Comparison of the effects of dark rearing and binocular suture on development and plasticity of cat visual cortex. *Brain Research*, 1981, *220*, 255–267.

Mower, G. D., Burchfiel, J. L., & Duffy, F. H. Animal models of strabismic amblyopia: Physiological studies of visual cortex and the lateral geniculate nucleus. *Developmental Brain Research*, 1982, *5*, 311–327.

Mower, G. D., Caplan, C., Burchfiel, J., & Duffy, F. H. Naloxone improves spatial resolution in MD cats. *Investigative Ophthalmology and Visual Science* (Supplement), 1982, *22*(3), 236.

Myslivecek, J., & Stipek, S. Effects of early visual and complex stimulation on learning, brain biochemistry, and electrophysiology. *Experimental Brain Research*, 1979, *36*, 343–357.

Nicoll, R. A., Siggins, G. R., Ling, N., Bloom, F. E., & Guillemin, R. Neuronal actions of endorphins and enkephalins among brain regions—Comparative microiontophoretic study. *Proceedings of the National Academy of Sciences, of the United States of America*, 1977, *74*, 2584–2588.

Nieto-Sampedro, M., Lewis, E. R., Cotman, C. W., Manthorpe, M., Skaper, S. D., Barbin, G., Longo, F., and Varon, S. Brain injury causes a time-dependent increase in neuronotrophic activity at the lesion site. *Science*, 1982, *217*, 860–861.

Nottebohm, F. A brain for all seasons: Cyclical anatomical changes in song control nuclei of the canary brain. *Science*, 1981, *214*, 18.

Palmer, D. H., Morris, D. A., Taylor, J. M., Stewart, B. J., & Hoffer, B. J. Electrophysiological effects of enkephalin analogs in rat cortex. *Life Science*, 1978, *23*, 851–860.

Parker, C. R., Simpson, E. R., Bilheimer, D. W., Leneno, K., Carr, B. R., & MacDonald, P. C. Inverse relation between low-density lipoprotein cholesterol and dehydroisoandrosterone sulfate in human fetal plasma. *Science*, 1980, *208*, 512–514.

Pettigrew, J. D., & Daniels, J. D. Gamma-aminobutyric acid antagomism in visual cortex: Different effects on simple, complex, and hypercomplex neurons. *Science*, 1973, *182*, 81–83.

Por, S., Bennett, E., & Bondy, S. Environmental enrichment and neurotransmitter receptors. *Behavioral and Neurological Biology*, 1982, *34*, 132–140.

Purves, D., & Lichtman, J. W. Elimination of synapses in the developing nervous system. *Science*, 1980, *210*, 153–157.

Rakic, P. Personal communication, 1984.

Rose, S. Lags in the cognitive competence of prematurely born infants. In S. L. Friedman & M. Sigman (Eds.), *Preterm birth and psychological development*. New York: Academic Press, 1981.

Rosenzweig, M. R., Krech, D., Bennett, E. L., & Diamond, M. C. Effects of environmental complexity and training on brain chemistry and anatomy. *Journal of Comparative and Physiological Psychology*, 1962, *55*, 429–437.

Schwartz, S. Effects on neonatal cortical lesions and early environmental factors on adult rat behavior. *Journal of Comparative and Physiological Psychology*, 1964, *57*, 72–77.

Shatz, C. J., & Stryker, M. P. Ocular dominance in layer IV of cats' visual cortex and effects of monocular deprivation. *Journal of Physiology* (London), August 1978, *281*, p. 267.

Sillito, A. M. The effectiveness of bicuculline as an antagonist of GABA and visually evoked inhibition in the cat's striate cortex. *Journal of Physiology*, 1975, *250*, 287–304. (a)

Sillito, A. M. The contribution of inhibitory mechanisms to the receptive field properties of neurones in the striate cortex of the cat. *Journal of Physiology*, 1975, *250*, 305–329. (b)

Sillito, A. M., Kemp, J. A., & Blakemore, C. The role of GABAergic inhibition in the cortical effects of monocular deprivation. *Nature*, 1981, *291*, 318–320.

Simantov, R., Kuhar, M. J., Pasternak, G. W., & Snyder, S. H. Regional distribution of a morphine-like factor enkepalin in monkey brain. *Brain Research*, 1976, *106*, 189–197.

Smith, D. Developmental alterations in binocular competitive interactions and visual acuity in visually deprived cats. *Journal of Comparative Neurology*, 1981, *198*, 667–676.

Smith, E. L., Bennett, M. J., Harwerth, R. S., & Crawford, M. L. J. Binocularity in kittens reared with optically induced squint. *Science*, 1979, *204*, 857–877.

Spinelli, D. N., & Jensen, F. E. Plasticity: The mirror of experience. *Science*, 1979, *203*, 75–78.

Spinelli, D. N., Jensen, F. E., & DePrisco, G. V. Early experience effect on dendritic branching in normally reared kittens. *Experimental Neurology*, 1980, *68*(1), 1–11.

Stone, J., & Dreher, B. Parallel processing of information in the visual pathways, a general principle of sensory coding? *Trends in Neuroscience*, 1982, *5*, 441–446.

Stryker, M. P., Sherk, H., Leventhal, A. G., & Hirsch, H. V. B. Physiological consequences for cat's visual cortex of effectively restricted early visual experience with oriented contours. *Journal of Neurophysiology*, 1978, *41*(4), 896–909.

Uphouse, L. L., & Bonner, J. Preliminary evidence for the effects of environmental complexity on hybridization of rat brain RNA to rat unique DNA. *Developmental Psychobiology*, 1975, *8*, 171–178.

Van Sluyters, R. C., & Blakemore, C. Experimental creation of unusual neuronal properties in visual cortex of kitten. *Nature*, 1973, *246*, 506–508.

Von Noorden, G. K., & Burian, H. M. Visual acuity in normal and amblyopic patients under reduced illumination: I. Behavior of visual acuity with and without neutral density filter. *Archives of Ophthalmology*, 1959, *61*, 533–535. (a)

Von Noorden, G. K., & Burian, H. M. Visual acuity in normal and amblyopic patients under reduced illumination: II. The visual acuity at various levels of illumination. *Archives of Ophthalmology*, 1959, *62*, 396–399. (b)

Von Noorden, G. K., & Burian, H. M. Perceptual blanking in normal and amblyopic eyes. *Archives of Ophthalmology*, 1960, *64*, 817–822.

Von Noorden, G. K. Reaction time in normal and amblyopic eyes. *Archives of Ophthalmology*, 1961, *66*, 695–701.

Von Noorden, G. K., & Leffler, M. B. Visual acuity in strabismic amblyopia under monocular and binocular conditions. *Archives of Ophthalmology*, 1966, *76*, 172–177.

Vorhees, C. V. Effects of prenatal naloxone exposure on postnatal behavioral development of rats. *Neurobehavioral Toxicology and Teratology*, 1981, *3*, 295–301.

Vrensen, G., & Cardozo, J. N. Changes in size and shape of synaptic connections after visual training: An ultrastructural approach of synaptic plasticity. *Brain Research*, 1981, *218*, 79–97.

Wesa, J. M., Chang, F. F., Greenough, W. T., & West, R. W. Synaptic contact curvature: Effects of differential rearing on rat occipital cortex. *Developmental Brain Research*, 1982, *4*, 253–257.

Wiesel, T. N., & Hubel, D. H. Receptive fields of cells in striate cortex of very young, visually inexperienced kittens. *Journal of Neurophysiology*, 1963, *26*, 994–1002. (a)

Wiesel, T. N., & Hubel, D. H. Simple cell responses in striate cortex of kittens deprived of vision in one eye. *Journal of Neurophysiology*, 1963, *26*, 1003–1017. (b)

Wiesel, T. N., & Hubel, D. H. Comparison of the effects of unilateral and bilateral eye closure on cortical unit responses in kittens. *Journal of Neurophysiology*, 1965, *28*, 1029–1040.

Will, B. E., Rosenzweig, M. R., Bennett, E. L., Herbert, M., & Morimoto, H. Relatively brief environmental enrichment aids recovery of learning capacity and alters brain measures after postweaning brain lesions in rats. *Journal of Comparative and Physiological Psychology*, 1977, *91*, 33–50.

Williams, J. T., & Zieglgansberger, W. Neurons in the frontal cortex of the rat carry multiple opiate receptors. *Brain Research*, 1981, *226*, 304–308.

Wise, S. P., & Herkenham, M. Opiate receptor distribution in the cerebral cortex of the rhesus monkey. *Science*, 1982, *218*, 387–389.

Yang, H. Y., Hong, J. S., & Costa, E. Regional distribution of len and met enkephalin in rat brain. *Neuropharmacology*, 1977, *16*, 303–307.

4

Discovering Messages in the Medium

Speech Perception and the Prelinguistic Infant

CATHERINE T. BEST

To parents and those who work with infants, one of the most remarkable developments during the first year of life is the rapid growth of vocal communication skills prior to language. Initially, the infant communicates by cries, but during the second half-year, the speechlike sounds of babbling emerge. By twelve months, babbling not only conveys feelings and needs to others but may also express infants' observations of regularities in the events and objects of their world.

The sounds that infants make are but one facet of their progress toward verbal communication. More hidden from our view, yet also important, is their *perceptual* grasp of the speech around them. In this chapter on infant speech perception, speech will be considered as the medium through which language is expressed vocally, much like the sounding of musical instruments is the medium through which a symphony is expressed. Both language and symphony are structurally complex systems, with many levels of concurrent organization, which are reflected in the organization of the medium. Speech carries the multiple messages that can be conveyed verbally, and hence carries information about the complex structural organization of vocal communication, which includes not only the structure of words and sentences but also broader aspects, such as stress, conversational rhythm, and voice

Catherine T. Best • Department of Human Development, Cognition, and Learning, Teachers College/Columbia University, New York, New York 10027. Work on the chapter was supported in part by a Spencer Young Faculty grant and Biomedical Research grant 42-USC 241 awarded through Teachers College.

characteristics (e.g., speaker gender, age, identity, and emotional state). It also carries information about finer grained structures such as consonants, vowels, and syllables, and the vocal gestures that produce them.

A listener's knowledge about the organization of vocal communication sets limits on which messages can be recognized in speech. That is to say, one must know or learn which structures to listen for within the wealth of information that the medium reflects about the events forming a vocalization (stated in the spirit of E. J. Gibson, 1977, and J. J. Gibson, 1966). Although infants may recognize some aspects of nonlinguistic structures in vocal communication, they are limited in their recognition of language structure as such. To use language, then, they must still discover the existence and meaning of many of the messages in speech, particularly those of words and phrases. But where do they start and how do they proceed in their discoveries during the first year? For infants who have not yet discovered the word, what messages are perceptually available to them in speech that might expedite that discovery?

The central concern of this chapter is with the nature of messages that prelinguistic infants may hear in human speech, at the level of the finer grained structures that we know as consonants and vowels. This issue has been addressed in the last twelve years of research on infant speech perception. Two basic themes about the information that infants perceive in speech have emerged. Both themes presume that some innate mechanism(s) of the auditory perceptual system fully account for infant speech perception, thus implying a mechanistic view that the young perceiver's role in seeking information in speech is rather passive. According to the first theme, infants possess species-specialized perceptual mechanisms that are tuned to linguistic contrasts among phonemes, those individual consonants and vowels we adults often associate with letters or letter combinations in words. The other theme proposes that infant speech perception is shaped by the auditory system's response to acoustic components of the stimulus; that is, the perceptual process is stimulus-bound and intrinsically neutral with respect to speech versus other sounds. These neutral acoustic attributes include the bits of noise and frequency changes, interspersed with silent gaps and humming or buzzing, which comprise a physical description of the speech signal.

It will be argued here that neither theme adequately explains how infants perceive the speech they normally hear during development. In their stead, the features of a third perspective based on ecological considerations (see also Fowler, Rubin, Remez, & Turvey, 1980; Summerfield, 1978) will be outlined, which posits a more active, information-seeking role for the perceiver. This alternative view is that infants actively attend for information in the speech medium about the natural forces that structured it, particularly how it was shaped by the human vocal tract. For the sake of simplicity, this perceptual focus on how speech is structured by its vocal source will be referred to as *speech source perception*. This term is offered rather than *articulatory perception* (see Studdert-Kennedy, 1981a; Summerfield, 1978), in order to encompass not only the articulatory gestures of the mouth and tongue but also the anatomical structure of the human vocal tract and its variations according to speaker characteristics (e.g., sex, age, and emotional state). As will be argued, a theory focused on the vocal tract sources of the speech medium's acoustic structure has greater potential than the other two themes for explaining how and why infants might begin to develop language based on the speech they hear. In short, it would provide the infant a more direct avenue by which to both discover and produce words.

But how is this vocal source information conveyed in speech? Simply stated, for now, the acoustic properties of sounds are determined by the structure and movements of the sound-making object, including the human vocal tract (Fant, 1960; Flanagan, 1973). Thus, speech carries information about vocal configurations and gestures (Cooper, 1981; Dudley, 1940; Paget, 1930); the *speech source* view proposes that this vocal tract information is available to perception. This view will be explained in more detail, with support from recent research with adults and infants. It will also be suggested that the specialization of the human left cerebral hemisphere for language reflects an attunement to detect information in speech about the articulatory gestures of the speaker's vocal tract. The chapter concludes by noting that speech source perception is only one contribution to the infant's development of language. In order to discover words and develop language, infants must also learn about the broader aspects of language from the natural context in which speech occurs.

1. SETTING THE CONTEXT

1.1. Language and the Prelinguistic Infant

Prelinguistic infants, by definition, do not yet produce true words; that is, their vocalizations apparently do not refer to objects and events in the way that the words of the adult language community do. It should be noted that it is not possible to draw a sharp chronological division between prelinguistic and linguistic periods in the development of either speech production or speech perception. Generally, however, the first year of life is considered to be prelinguistic.

Two complementary and interdependent questions provide a guide for understanding the prelinguistic antecedents of language development: What is vocal language that a prelinguistic infant may come to know it? and What is the prelinguistic infant that she or he may come to know vocal language? (adapted from McCulloch, 1965). The next few sections will focus on the former question, to frame the subsequent discussion of infant speech perception research. They will describe the basic characteristics of speech that are important for understanding the task facing a prelinguistic perceiver. Once the stage has been set, the three theoretical views about the way infants process speech will be described in greater depth.

1.2. What Is Vocal Language?

What type of information or messages does speech carry that prelinguistic infants might perceive? Prelinguistic infants do not yet produce words, nor do most infants under 9–10 months yet comprehend spoken words (e.g., Lenneberg, 1967). Thus, we should not expect younger infants to perceive any information that is defined by word meanings. Nevertheless, some coherent information in human speech must be available to prelinguistic infants, for they do eventually discover words.

To discover words in the speech directed toward them, presumably infants would, in part, have to (a) disembed from continuous speech the recurring subpatterns that become familiar words; (b) recognize the invariance in the pattern of a word, across the variations in acoustic detail that occur when it is produced by different speakers or in different contexts; and (c) recognize the relevant differences that do specify meaningfully different patterns. And in order to produce words, they would also have to recognize how to imitate or approximate subpatterns from a language-user's speech, even though the acoustic output of their own smaller and differently proportioned vocal tracts differs substantially from that provided by their older models (Goldstein, 1979; Lieberman, Harris, Wolff, & Russell, 1971).

In language research with adults, phonemes (consonants and vowels) have often been considered to be the building blocks of words. According to that perspective, the achievement of the perceptual tasks previously listed would seemingly be founded on perception at the phonemic level of speech. To date, most infant speech perception research has focused on phonemes or their combinations in syllables, on the apparent assumption that perception of these subword units must be precursory to the perception of words.

Language users easily recognize words and phonemes when listening to conversational speech. However, these recognitions are no small feat for prelinguistic listeners, who lack a language system that could help them solve some apparent puzzles in adult speech perception. The source of these puzzles, which lies in the acoustic characteristics of speech, is discussed next. (For more extensive discussions of adult speech perception, see Fowler *et al.*,

1980; Liberman, 1982; Liberman, Cooper, Shankweiler, & Studdert-Kennedy, 1967; Pisoni, 1978).

1.3. The Puzzles in the Acoustic Shape of Speech

To imagine the infant's difficulty, recall listening to a stretch of conversation in an unfamiliar language. Foreign speech typically seems like a relatively continuous flow of sounds, in which the quality of some phonemes may be unfamiliar (e.g., the /r/ of French or Spanish), and the boundaries of individual words often may be indecipherable. This can make it difficult for a listener to recognize a foreign word uttered in different sentences and by different people. The infant's problem is compounded because, in contrast to a language user, infants presumably do not know what words are, so this concept cannot guide their discovery of word boundaries in the flow of conversational speech. Similarly, a language user may have difficulty recognizing the precise qualities of an unfamiliar foreign phoneme when it is uttered in different words and by different people. Yet, relative to mature language users, to infants all of the phonemes occurring even in their native language environment would be comparatively unfamiliar. Infants also presumably lack certain concepts about the linguistic role of phonemes that may guide the language user's recognition of individual phonemes in conversational speech.

1.3.1. Acoustic Continuities and Discontinuities in Running Speech

One reason a sentence spoken in an unfamiliar language sounds indivisible is that utterances in natural conversation are a fairly continuous stream of sound. This is partly attributable to the cohesive intonation, or pitch contour of the voice. But it results also from the vocal-tract movement trajectories that interconnect the adjacent words in sentences or phrases. Speakers in conversation rarely pause between words, instead usually moving in connected fashion from one to the next, just as a runner usually adjusts to changes in terrain or direction without pausing between step cycles. Sometimes neighboring words in informal speech even become contracted (e.g., "what are you . . ." becomes "wadaya . . ." or "whatcha . . ."). Thus, the raw acoustic properties of conversational speech do not always reveal clear boundaries between words.

Conversely, the vocal tract can make other relatively rapid adjustments that do cause obvious acoustic discontinuities. These breaks can occur within as well as between words, however. For instance, in the word "so" there is a rather sudden change from the lack of vocal-cord vibration during the voiceless /s/ to the onset of vibration for the voiced sound /o/. This causes an acoustic break between the noiselike, aperiodic hiss of the /s/ and the voiced acoustic periodicity of the vowel. The paradox is that a knowledgeable listener perceives these discontinuities as an integral part of a word, where appropriate, rather than as breaks between words.

The speech properties just discussed can be seen in Figure 1. A spectrogram is one way of visualizing the acoustic components of speech. As indicated earlier, these acoustic characteristics are determined by the structure and movements of the vocal tract (Fant, 1960; Flanagan, 1973).

The spectrographic analysis shows the relative acoustic intensity (darkness level) of the frequency components in the speech signal (ordinate) as they change over time (abscissa). The wide, horizontally varying bars of increased density within the dark vertical striations are called *formants*, the lowest being referred to as the first format (F1), and correspond to the time-varying resonant frequencies of various relatively hollow spaces or chambers in the vocal tract. In the vowel "ee," for example, a small resonating chamber is formed at the front of the mouth between the edges of the tongue blade pressed against the upper teeth and the close approach of the soft palate to the base of the tongue, while a relatively large resonating chamber forms at the back of the mouth behind the base of the tongue. This results in a low-frequency F1 and a high-frequency second format (F2).

The vertical striations in which the formats appear represent the individual energy pulses emitted by each vocal-fold vibration. The more closely packed the striations are, the briefer the periods between pulses and hence the higher the pitch of the voice. In Figure 1, the man raised his voice pitch substantially for the word "saying" and then dropped it for "(to) me little," raising it again toward the end of "girl." This degree of pitch modulation is more exaggerated than normal, and often occurs when parents talk playfully to their babies (Kaye, 1980). The dappled, nonstriated patches represent aperiodic acoustic noise produced by air turbulence at some point in the vocal tract as with the tongue-tip constriction near the upper front teeth for the /s/ of "saying."

1.3.2. Continuities and Discontinuities among and within Phonemes

Since infant speech research has focused primarily on phonemes, the sentence in Figure 1 is printed beneath the spectrogram for reference. The letters for the vowels and consonants are roughly lined up under the midpoint of their portion of the acoustic signal. The match between phoneme and acoustic information is only approximate, however, because of inherent difficulties in determining the acoustic span of a phoneme. At times, discontinuities appear to fall within rather than between phonemes, as was true at the level of words in a sentence. For instance, the /t/ in "to" encompasses

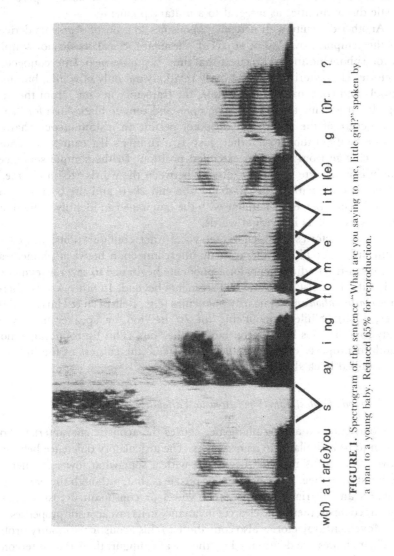

FIGURE 1. Spectrogram of the sentence "What are you saying to me, little girl?" spoken by a man to a young baby. Reduced 65% for reproduction.

an aperiodic noise burst, the following brief period of breathy aspiration, and the subsequent rapid transitions of the format frequencies (i.e., the upglide at onset of the lowest formant and downglides in the pale higher formants). Again, the knowledgeable listener may paradoxically perceive acoustic discontinuities as integral to a unitary phoneme.

Another difficulty with matching phonemes to acoustic segments derives from the temporal overlap of adjacent phonemes. Speakers do not simply finish one phoneme and, at precisely that time, begin the next. Interconnected trajectories characterize the neighborly relations not only of words, but also of vowels and consonants, for example, the trajectory in "me" from the lips being closed for /m/ to the lips being open and tongue blade high for "ee." The structure of the vocal tract does not permit an instantaneous change from one configuration to another, just as a runner's leg cannot instantaneously change from flexed to extended position. In the sample sentence, not only is a discrete boundary missing between the words "to" and "me," but there is none within "to" to define the end of /t/ and the beginning of the vowel. Yet a perceiver familiar with the language can identify individual phonemes as well as individual words.

The trajectories between target vocal-tract configurations, however, account only partially for the acoustic interconnection between phonemes. Often, vocal-tract adjustments for a phoneme begin one to several segments ahead of it, or persist one to several segments beyond. In other words, there is some *coarticulation* among nearby phonemes (e.g., Bell-Berti & Harris, 1979; Fowler, 1980). While pronouncing the /t/ in "too," a speaker usually is already rounding his lips appropriate for the "oo." This lip-rounding is not a standard property of /t/—for the /t/ in "tee," the corners of the lips are instead pulled back slightly for the following "ee."

1.3.3. Phonemic Context Effects and Acoustic Variability

Coarticulation among phonemes causes the acoustic characteristics of any item to be assimilated to its neighbors. The articulatory difference between the two /t/'s results in "too" beginning with a somewhat lower frequency noise burst than "tee." The paradox or puzzle is that, although the perceiver recognizes an invariant identity for a vowel or consonant across various phonemic contexts, there is no clearcut invariance in its raw acoustic properties.

Movement trajectories also contribute to this acoustic variability problem. Their shapes are determined by the vocal configurations they interconnect and there are rarely definable boundaries in conversational speech between a static configuration and a trajectory into or out of it. Figure 1 indicates that, because of the differences in the surrounding phonemes, the first and the last /l/ in "little" are acoustically different, both in the flanking format trajectories and in the exact frequencies of the flatter formants midway through the "segment."

1.3.4. Vocal Tract Variations and Perceptual Normalization

Not illustrated in the figure is a broader problem of acoustic variability: the acoustic contextual variation caused by different speakers. Of importance are the differences found between males and females, or between children and adults. On the average, female vocal tracts are smaller than those of males, which biases the acoustics of female speech toward higher frequencies in voice pitch and in formant frequencies. More important, though, are the age and gender differences in proportional relations among vocal-tract areas. The ratio between the distance from the vocal cords to the base of the tongue, versus the distance from the lips to the base of the tongue, is greatest for adult males and smallest for young infants (Goldstein, 1979; Lieberman, Harris, Wolff, & Russell, 1971).

Because formant frequencies are determined by the sizes of the vocal resonating chambers, these vocal-tract ratio differences cause age and gender differences in the proportional relations among formant frequencies for a given vowel. It has been impossible thus far to derive a simple mathematical formula for the formant frequency relations of a vowel produced by proportionally differing vocal tracts. In other words, there is no invariant acoustic description of formant frequency relations across men's, women's, and children's utterances of the vowel (Bernstein, 1981; Broad, 1981; Kent & Forner, 1979). The puzzle is that listeners, at least those familiar with the language, immediately hear the vowel's identity across a variety of vocal tracts differing in size and proportion. This perception of constancy in the face of speaker-specific acoustic variations has been referred to as the vocal-tract *normalization* problem.

1.3.5. Summary

These acoustic properties thus pose a number of difficulties for the perceptual capture of words or phonemes from conversational speech, even for adults listening to their native language. These difficulties can cause a sentence spoken in an unfamiliar language to sound like a rather undivided flow, when the listener is not prepared to handle them. However, when adults listen to their native language, they can identify discrete phonemes and words. Most likely, this is because they already know the phonemes and many of the words in their language, as well as the permissible ways by which items of either type can combine. Infants, on the other hand, do not have this knowledge of language. Yet they must be able to "solve these puzzles" in perceiving speech in order to ultimately discover words, since the words directed to infants are usually embedded in phrases or sentences (Kaye, 1980) and are presented to them by a variety of people in different speech contexts. What might infants perceive in speech, at the level of the phoneme, that could help them to recognize discrete words within the flow of sound? For consideration of this question, the discussion now turns to the infant speech perception literature.

2. THE FOUNDATIONS OF INFANT SPEECH PERCEPTION RESEARCH

Research on infant speech perception has largely been guided by two theoretical approaches to one overriding issue and its underlying assumptions, as indicated earlier. Although the questions and discussion presented thus far have been oriented around the eventual discovery of words by infants who initially have very limited knowledge about vocal communication, this has not been the major issue in research and theory on infant speech perception. Instead, the primary theoretical focus has been on how infants solve the acoustic puzzles of phoneme perception (e.g., see Jusczyk, 1981a,b). Its main underlying assumptions have been that (a) the basic perceptual unit in speech *is* phoneme-sized, (b) the speech percept derives from intraperceiver transformation(s) of the acoustic properties of the stimulus, and (c) the source of the transformation is an innate mechanism of the auditory system.

Much of the research has been generated by a controversy over the nature of the transformation(s) and supporting mechanism that can be traced to a similar controversy in experimental work on adult speech perception. On one side is the *phonetic* interpretation of infant speech perception, which posits that the perceptual mechanism is uniquely human by nature and differs qualitatively from the means for perceiving other sounds. Proposals about the exact properties of the specialized phonetic mechanism have ranged from a comparator that matches incoming speech sounds to the neuromotor commands for producing them (the motor theory of speech perception: Liberman *et al.*, 1967) to innate categories of linguistic features of phonemes (e.g., Eimas, Siqueland, Jusczyk, & Vigorito, 1971) that may be mediated by innately tuned neural feature-detectors (e.g., Cutting & Eimas, 1975).

On the other side of the controversy is the *psychoacoustic* approach, which presumes the machinery for the speech-to-percept transformation to be neither uniquely human nor limited to the perception of speech. In the psychoacoustic view, the general organization of the mammalian (or primate) auditory system yields an invariant stimulus-bound response whenever a given acoustic property occurs, regardless of the class of sound (e.g., speech vs. nonspeech) to which the individual is listening.

In the following summary and interpretation of the literature, it will be argued that the psychoacoustic-phonetic controversy in infant speech perception research is misguided. Both views are inadequate because they fail to consider the relation between infant and language. Following that review and discussion, a promising alternative theoretical perspective on infant speech perception will be described: the ecologically motivated (e.g., J. J. Gibson, 1966; Summerfield, 1978) *speech source perception* view outlined earlier. But at this point, the issue that has guided existing infant speech perception research, the psychoacoustic-phonetic controversy, must be placed in proper historical perspective.

2.1. The Empirical Beginnings

Research on infant perception of phonemes began with two reports in 1971, both of which gave a phonetic interpretation to the underlying processes. Each study employed a variant of the habituation paradigm to map the limits of infants' phoneme categories via their discriminations of syllable pairs differing in initial consonant. One of the studies found that 5 to 6-month-olds can discriminate natural utterances of /ba/ and /ga/ (Moffitt, 1971). The consonants in the tested syllables are both voiced stop consonants (along with /d/; the voiceless stops are /p/, /t/, and /k/), but they differ in place of articulation. Since the infants discriminated between consonants that differed solely in place of articulation, the author concluded that "linguistic-perceptual capacities are present during early life" (p. 717).

The other study (Eimas et al., 1971) has received the preponderant attention in subsequent research. In this study, 1- and 4-month-olds were presented with computer-synthesized versions of /ba/ and /pa/, which differ in the articulatory property called *voice onset time* (VOT). That is, in the voiceless /p/ of /pa/, the vocal cords begin vibrating later with respect to the lip-opening gesture than is the case with the voiced /b/ of /ba/. Therefore, /p/ has a longer VOT than /b/. The acoustic consequences of articulatory differences in VOT are many (Lisker, 1978), but research attention has primarily focused on the time difference between the consonantal noise burst and the onset of periodic voicing, referred to here as *acoustic* VOT. It is usually confounded with other acoustic differences between a voiced-voiceless consonant pair.

Computer synthesis was used in the Eimas et al. study to produce a systematic series or continuum of syllables, which varied in equal-sized steps along the acoustic VOT dimension. Such acoustically controlled continua usually cannot be produced by a human speaker because of mechanical constraints on possible vocal tract movements (although in the case of stop voicing, human speakers can produce a range of different acoustic VOT values). Adult listeners typically fail to hear the gradual steps of acoustic change along synthetic continua between two contrasting consonants. Instead, they identify all stimuli as exemplars of one or the other phoneme category, and a sharp boundary on the continuum separates the two perceptual categories. Adult listeners also discriminate between acoustically different pairs of synthetic stimuli much better when the members are from different phoneme categories than from the same category. This pattern of identification and discrimination results has been termed *categorical perception* (see Figure 2). It was originally taken as evidence for a specialized phonetic mode of perception, since the nonspeech continua that had been similarly studied were perceived continuously (i.e., no clear labelling boundary or no clear performance peak in discrimination ability) rather than categorically (e.g., Liberman *et al.*, 1967; Repp, 1982).

To determine whether infants also perceive speech categorically, and by presumption phonetically, Eimas *et al* (1971) tested their discrimination of synthetic syllables that differed in acoustic VOT, but either did or did

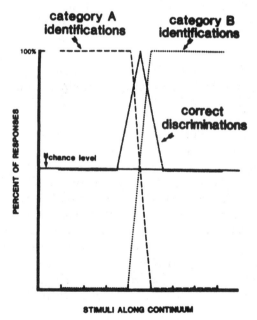

CATEGORICAL PERCEPTION

FIGURE 2. A schematic diagram of ideal results of categorical perception tests.

not differ according to American English /pa/ versus /ba/. The within-category pairs differed by the same magnitude of acoustic VOT as did the between-category pairs. The infants discriminated the between-category difference much better than the within-category difference, in agreement with the adult phoneme boundary. This finding led the authors to conclude that infants have an innate capacity to perceive speech linguistically, that is, in terms of adult phoneme categories.

These two early studies were scientifically intriguing, and encouraged a profusion of infant speech research that still continues. Some researchers essentially accepted the premise that the means by which acoustic properties are translated to a percept is speech-specialized, that is, specifically phonetic in nature. They went on to explore the range and nature of phoneme contrasts that young infants discriminate. As the psychoacoustic-phonetic controversy indicates, however, other researchers took issue with the phonetic perspective. The studies of this latter group have attempted to show that categorical perception is due to general psychoacoustic mechanisms, which respond to particular acoustic attributes whether they appear in speech or nonspeech sounds, but which occur most frequently in speech (e.g., Blumstein, 1980; Stevens & Blumstein, 1978).

2.2. Infant Perception of Phoneme Contrasts

Many of the early studies employed synthetic syllable continua, since explaining categorical perception has been a key interest in the psycho-acoustic-phonetic controversy. According to the definition of categorical perception, percentage-correct discrimination of stimuli from a continuum must show a performance peak at the position of the category boundary for labeling of those stimuli (see Figure 2) (Studdert-Kennedy, Liberman, Harris, & Cooper, 1970). Therefore, both identification and discrimination data are needed to assess categorical perception. However, because there is no currently acceptable test for infants' identification of sounds, the conclusions of the infant research are based on discrimination data only (see Jusczyk, 1981a). Since the infant discrimination data cannot be compared to infant identification responses, it may be better to refer to the peaks and troughs in their discrimination of synthetic speech continua by some other term, such as perceptual *boundary effect* (Kuhl, 1981b; Wood, 1976).

A number of other infant studies from the phonetic perspective have employed natural rather than synthetic speech stimuli, thus focusing on discrimination of contrastive spoken examplars rather than on categorical discrimination. In both types of study, the discrimination data have been derived from either habituation tests or tests of generalization of conditioned operant responses, and the subjects have usually been between 1 and 7 months of age (prior to the reported onset of word comprehension). The consonant contrasts that were tested have usually occurred in syllable-initial position, which makes phoneme discrimination in natural speech easier for young children than do medial or final positions (Schvachkin, 1973).

The categorical discrimination studies have found that infants show a boundary effect for a variety of synthetic consonant contrasts, with their discrimination peak usually occurring at or near the adult American English identification boundary. These contrasts include stop consonant voicing (specifically, /p/ vs. /b/, and /d/ vs. /t/) as cued by acoustic VOT (Eimas, 1975a; Eimas et al., 1971; Streeter, 1976) or by another naturally occurring cue, the extent of frequency change in the first formant (F1) transition (J. L. Miller & Eimas, 1981).

Infants also show a boundary effect for place of articulation differences between stop consonants (/b/ vs. /d/ vs. /g/). (C. L. Miller & Morse, 1976; C. L. Miller, Morse, & Dorman, 1977; Morse, 1972; Williams & Bush, 1978), even when the consonants occur in the middle of vowel-consonant-vowel (VCV) syllables (Jusczyk & Thompson, 1978) or in the final position of VC or CVC syllables (Jusczyk, 1977). Infant boundary effects have also been found for place of articulation distinctions between liquid consonants (/r/ vs. /l/, Eimas, 1975b). There is an infant boundary effect for place of articulation differences among fricatives (voiceless: /f/ vs. "th" as in "thanks"; voiced: /v/ vs. "th" as in "that"), although the infants' discriminations may differ in some respects from adults' (Jusczyk, Murray, & Bayly, 1979). Infants also show a boundary effect for manner of articulation differences between /b/–/w/ (same place of articulation, but stop vs. semivowel manner: Eimas & Miller, 1980a; Hillenbrand, Minifie & Edwards, 1979) and between /b/ –/m/, although they do not discriminate the latter contrast as categorically as adults, that is, they show moderate discrimination of the within-category pairs (Eimas & Miller, 1980b).

Adults perceive synthetic continua between vowel contrasts in a more continuous or less categorical fashion than consonants, unless the vowels are severely shortened in duration (e.g., Crowder, 1973; Liberman et al., 1967; Pisoni, 1973, 1975). Infants show similar effects in perception of the "ee-ih" vowel distinction (Swoboda, Morse, & Leavitt; Swoboda et al., 1978).

Consistent with the boundary effect findings, infants discriminate a fairly wide array of the natural consonant and vowel contrasts in spoken English. They discriminate natural stop voicing distinctions (Trehub & Rabinovitch, 1972), as well as some contrasts that children often do not produce correctly until late in phonological development (3–5 years), and which often cause persistent articulatory difficulties: certain fricative place (/s/–"sh") and voicing (/s/–/v/) contrasts (Eilers & Minifie, 1975; Eilers, Wilson, & Moore, 1977), the place contrast between the liquid consonants

/w/–/r/ (Eilers, Oller, & Gavin, 1978), and the consonant clusters /sl/–/spl/ (Morse, Eilers, & Gavin, 1982). Infants also discriminate the naturally produced vowels "ee" versus "ih" (Eilers *et al.*, 1978), as well as the "ee-ah-oo" triad, whether they occur in isolation or in CV syllables (Trehub, 1973).

To summarize, young prelinguistic infants discriminate many consonant contrasts and often show a boundary effect akin to the adult category boundary. They even discriminate some consonant contrasts that children produce late and often misarticulate. Infants also discriminate vowel contrasts, and do so less categorically than consonants, again like adults. (For more extensive discussion of the theoretical particulars, see Aslin & Pisoni, 1980; Cutting & Eimas, 1975; Eilers, 1980; Eilers & Gavin, 1981; Eimas, 1974a, 1975a; Jusczyk, 1981a, 1981b; Kuhl, 1978, 1980, 1981a; Mehler & Bertoncini, 1978; Morse, 1978; Trehub, Bull, & Schneider, 1981; and Walley, Pisoni, & Aslin, 1981.)

The performance pattern does not, however, indicate whether infants discriminate by psychoacoustic or phonetic means. Initially, researchers who took a phonetic view assumed that the boundary effect was evidence for a speech-specialized perceptual process. But that assumption could be questioned, and was submitted to test by researchers on both sides of the theoretical dichotomy. The alternative posed by the psychoacoustic perspective was, of course, that the perceptual boundary might be an attribute of the auditory system's response to the acoustic properties, rather than the phonemic identities, of the speech sounds.

2.3. Is the Boundary Effect "Phonetic" or "Psychoacoustic"?

A direct test of this question was to see whether infants show a discrimination boundary effect for some phoneme-differentiating acoustic property even when it occurs outside a speech context. Both Morse (1972) and Eimas (1974b, 1975b) isolated the major acoustic cue that had been manipulated to produce a place of articulation continuum, by stripping away the other acoustic properties that were shared by the contrasting phonemes, and presented young infants with the isolated cue continuum to discriminate. Morse (1972) tested infans with the isolated F2 for the /ba/–/ga/ contrast, whereas Eimas tested the isolated F2 cue for /da/–/ga/ (1974b), and the isolated F3 cue for /ra/–/la/ (1975b). Although in each case the isolated formants were the sole acoustic property that distinguished the phoneme contrast, and hence were crucial to adult categorical perception of that contrast, outside of their natural context they sounded like nonspeech "bleats."

The argument was that if the infant boundary effect reflects a uniquely speech-related perceptual specialization, it should occur in the perception of a particular acoustic cue only when that cue actually specifies a phoneme distinction. The infants in the Morse and Eimas control studies did show a boundary effect for the full syllables but failed to show one for the isolated formants, and in fact discriminated the latter very poorly. The authors interpreted their findings as support for the phonetically specialized nature of infant boundary effects.

The psychoacoustic perspective offered an alternate interpretation, however. The crucial stimulus attribute for psychoacoustically based boundaries could be the *interrelation* between the distinctive acoustic cue and the nondistinctive information provided in the other formants (e.g., Jusczyk, 1981a). Such a relational attribute would be destroyed by presenting the distinctive cue in isolation. Therefore, the Morse and Eimas studies could not definitively answer the controversy.

A more appropriate nonspeech control should maintain the interrelations among the acoustic features involved in a phoneme distinction. One such nonspeech distinction is the difference in risetime, or time from onset of a sound until it reaches its maximum intensity, between plucked versus bowed violinlike sounds, which is analogous to the "sha-cha" distinction in speech. Adults had been reported to perceive the pluck-bow distinction categorically, even though it is nonspeech (Cutting & Rosner, 1974). Jusczyk, Rosner, Cutting, Foard, and Smith (1977) extended that finding to infants and presented it as evidence that infant perceptual boundary effects have a psychoacoustic basis. However, subsequent replications of the adult study uncovered a stimulus problem. The acoustic differences among the original pluck-bow stimuli were not of equal magnitude throughout the continuum; when this source of acoustic discontinuity was removed, adults no longer perceived pluck-bow categorically (Rosen & Howell, 1981). Since Jusczyk *et al.* employed the original pluck-bow continuum, the infant findings must be questioned (Jusczyk, 1981a,b).

A second infant nonspeech study was subsequently run, using *tone onset time* (TOT) differences between the individual tones of a two-tone chord, which is an analogue for the acoustic VOT distinction in speech (Jusczyk, Pisoni, Walley, & Murray, 1980). Adults show a sharp TOT boundary in line with their boundary for acoustic VOT (Pisoni, 1977). Thus, the phonetic uniqueness of adult categorical perception has been called into question by the TOT results, along with similar reports on other nonspeech contrasts (e.g., J. D. Miller, Wier, Pastore, Kelly, & Dooling, 1976). In the Jusczyk *et al.* (1980) study, the infants discriminated TOT differences nearly categorically, leading the authors to conclude that earlier reported VOT boundary effects may not be unique to speech perception by infants either.

Nonetheless, the TOT findings also fail to offer a definitive choice between the phonetic and the psychoacoustic interpretations of infant perceptual boundaries. In contrast to adults, the infants failed to discriminate

TOT as categorically as acoustic VOT, and the position of their TOT boundary differed significantly from their acoustic VOT boundary. Jusczyk *et al.* (1980) claim that this does not damage the general psychoacoustic stand, since TOT only partly captures the *articulatory* VOT distinction (i.e., the psychoacoustic key could be some other, untested acoustic attribute of articulatory VOT). However, the developmental data are at odds with this logic. The TOT and acoustic VOT boundaries *do* match for adults, indicating that a perceptual change must occur between infancy and adulthood. Yet the infant and adult VOT boundaries match. Therefore, it is the TOT boundary that changes developmentally, and not the VOT boundary, in contradiction to the claim that infants come to perceive speech distinctions via psychoacoustic means (Jusczyk, 1981b).

In any event, the evidence from the infant nonspeech perception research is weak. The currently available nonspeech control studies present a theoretical stalemate between the phonetic and the psychoacoustic claims about the nature of infant speech perception. However, the proponents of the controversy have argued that the answer may lie elsewhere in the premises of the psychoacoustic-phonetic distinction. It might be settled by exploring whether phonetic perception is uniquely human, a claim made by the phonetic perspective that is rejected by the psychoacoustic perspective.

2.4. Is the Boundary Effect Uniquely Human?

Empirical attacks on the claim that humans are the sole possessors of categorical phoneme perception have involved assessing whether other mammals or primates show abrupt shifts in perceptual sensitivity around the phoneme boundaries. It was reasoned that, if animals showed a boundary effect, general psychoacoustic factors must then account for categoricity in speech perception, since by definition infrahumans cannot perceive in a humanly specialized manner. The relevance of this research to infant speech perception is that infants and animals are nonusers of language, but only infants are human and have the capacity to develop human language.

Researchers of animal speech perception have reported boundary effects, similar to those found with infants, for discrimination of the stop consonant voicing distinction by chinchillas (South American rodent) (Kuhl, 1981b; Kuhl & Miller, 1975; J. D. Miller, Henderson, Sullivan, & Rigden, 1978), and by rhesus monkeys (Waters & Wilson, 1976), and for the stop consonant place of articulation distinction by monkeys (Morse & Snowden, 1975; Sinnott, Beecher, Moody, & Stebbins, 1976). Chinchillas also discriminate the vowels "ah" and "ee" (Burdick & Miller, 1975); as do dogs (Baru, 1975), and exhibit boundary effects in go- no-go categorizations of voicing among stop consonants, with their boundaries falling at the position of human adult boundaries (Kuhl & Miller, 1978). Thus, the psychoacoustic interpretation is that the perceiver's knowledge of language is not a necessary precondition for the boundary effect, and apparently neither is membership in the human species.

Researchers who support the psychoacoustic view of speech perception have used three animal findings to propose the following picture of the evolution of speech perception and production: the mammalian auditory system has specialized notches in sensitivity for certain regions of certain acoustic dimensions. These psychoacoustic specializations placed selective pressures on the choice of phoneme contrasts by human languages. The productions of the phonemes chosen must have capitalized on just the acoustic domains that are most neatly suited to mammalian psychoacoustic specializations (Kuhl, 1981b; Kuhl & Miller, 1975, 1978; J. D. Miller, 1977; Stevens, 1972). By extension, human infants possess those same psychoacoustic sensitivities (Aslin & Pisoni, 1980; Jusczyk, 1981a,b; Kuhl, 1978; Walley, Pisoni, & Aslin, 1981), and their attention is thus captured by the acoustic attributes of the language in their environment.

However, the claims of this psychoacoustic proposal belie the clarity of the infrahuman data. The animal data fail in several ways to match those of human adults. Recall the earlier argument that a determination of categorical perception depends on data from labeling and from discrimination tests; the animal research has necessarily relied only on discrimination data. Indeed, animal discrimination boundaries are less sharply defined than human adult phoneme boundaries (Kuhl & Miller, 1978). In other words, animals are noticeably better than human adults at discriminating within-category acoustic differences in a place of articulation continuum, and worse at discriminating between-category differences, indicating lowered categoricity (Morse & Snowden, 1975). The between-species categoricity difference is consistent with the greater sensitivity of human adults to formant-onset frequency changes, by a factor of about two (Sinnot et al., 1976). In addition, only human adults show reaction time increases, large ones in fact, when making within-category discriminations (Sinnott et al., 1976). Finally, the absolute position of the human adult category boundary is more stable than the monkey boundary, in the face of variations in the acoustic range covered by a synthetic phoneme continuum (Waters & Wilson, 1976). Humans obviously do show speech-relevant perceptual specializations beyond the limits of the other mammals tested.

The animal and nonspeech research may suggest that the boundary effect is not absolutely speech-specific and species-specific. However, there are clear and unexplained differences remaining between the human adults' perception of speech and the control studies on animal speech perception and infant nonspeech perception. Thus, the basic theoretical choice between the phonetic and psychoacoustic explanations of speech perception, especially in infants, is still open. If studies of the boundary effect have failed to solve the psychoacoustic-phonetic quandary, then possibly a more abstract characteristic of speech perception would (such as the perceptual constancy of phoneme identity across phonemically irrelevant acoustic variations).

2.5. Phonemic Perceptual Constancy in Infants

Recall the earlier discussion of some puzzles in the fit between speech acoustics and perception, particularly the lack of satisfactory acoustic descriptions for the invariant identity of a phoneme across different contexts of surrounding phonemes (the acoustic variability problem) or as uttered by differently proportioned vocal tracts (the normalization problem). In spite of these puzzles, adults perceive the identity of a phoneme spoken in widely different words and by different vocal tracts with seeming immediacy and effortlessness.

To see whether infants show a similar perceptual constancy, Fodor and colleagues (Fodor, Garrett, & Shapero, 1970; Fodor, Garrett, & Brill, 1975) trained them to respond operantly to a pair of vowel-differing syllables that either began with the same consonant (e.g., "pee"–"poo")or began with different consonants (e.g., "pee"–"kah"). In both conditions, the consonants differed acoustically because of the change in vowel context. The authors wanted to assess whether, despite the acoustic variability, the infants learned the operant response more easily for the consonantal match. The infants were later tested on a new syllable (e.g., "pah"), to determine whether they generalized the learned operant response more consistently from the consonant-matched pair to a new syllable beginning with the same consonant. The infants did learn and generalize more consistently for consonantal matches than consonantal mismatches. If they had learned to associate syllable pairs simply by remembering the pairing of their dissimilar acoustic properties, they should have responded to the mismatched-consonant pairs as consistently as to the consonant-matched pairs. The authors concluded that these prelinguistic infants had maintained perceptual constancy for consonantal identity in the face of concurrent acoustic variations, and that this ability must depend "on innately determined phonological identities (p. 180)."

There are two problems with this claim. Their task was difficult for infants to learn, and several attempted replications or extensions have failed. In addition, the psychoacoustic perspective offered an alternative interpretation: The perceptual constancy might reflect a response to some (yet uncovered) higher order acoustic invariant shared by the varying instances of a given consonant (Kuhl, 1980, 1981b). Using a different technique than Fodor *et al.*, Kuhl and her colleagues found perceptual constancy in young infants for the vowels "ah" versus "ee" (Kuhl, 1979) and for the fricatives /f/ versus /s/ (Holmberg, Morgan, & Kuhl, 1977) across different neighboring phoneme contexts and different speakers. Thus, consistent with the Fodor *et al.* report, the infants solved the acoustic invariance problem. Since perceptual constancy was maintained across different speakers the infants also solved the normalization problem. But whereas Fodor *et al.* favored the phonetic viewpoint, Kuhl and others favor the psychoacoustic viewpoint (e.g., Jusczyk, 1981b; Walley *et al.*, 1981), in part because chinchillas and dogs show perceptual constancy for "ah" versus "ee" across speakers and pitch contours (Baru, 1975; Burdick & Miller, 1975). Chinchillas also show such constancy for /t/–/d/ even in different vowel contexts (Kuhl & Miller, 1975).

The perceptual constancy findings thus indicate that infants can somehow solve two seemingly knotty acoustic puzzles to reach an important perceptual aspect of phoneme identity. Once again, the findings apparently do not allow a theoretical choice between the phonetic and psychoacoustic explanations of infant speech perception. Also, as will be argued in the next section, neither is the theoretical choice decided by considerations about the innateness of infant phoneme perception.

2.6. Innateness of Infant Phoneme Perception Effects

A pervasive notion on both sides of the psychoacoustic-phonetic controversy has been that a boundary effect or perceptual constancy effect is innate if infants show it "at the earliest age tested" (e.g., Aslin & Pisoni, 1980; Eimas, 1975a; Eimas *et al.*, 1971; Jusczyk, 1981a,b; Kuhl, 1978). Curiously, the empirical foundation for this belief includes almost no data before 1 month, a handful of studies on 2-month-olds, and many studies that have collapsed data across 1–4 months, 4–6 months, 6–8 months, or 10–12 months. Few have compared different age groups (cf. Best, Hoffman, & Glanville, 1982; Eilers, Wilson, & Moore, 1977; Werker, 1983; Werker & Tees, 1982).

This view apparently assumes that "an infant is an infant is an infant," across at least the first 6 months of life. In studies that averaged over several months of age, the "earliest age" cannot be trusted since it refers to the youngest infant they tested, even though group data were reported and appropriate age analyses were almost never run (but see 1- versus 4-month-age differences in Eimas *et al.*, 1971). Thus, it is nearly impossible to assess which, if any, perceptual boundaries are innate, or presumably biologically determined and inborn. All we can note are the ages below which a given perceptual effect has *not* been shown; for this review, the conservative assumption will be that the average age tested is the correct "earliest age" to show the reported effect.

The literature offers the following observations: First, infant boundary effects matching the adult findings have been shown only by a mean of 2½ months (e.g., Eimas, 1974b, 1975b; Morse, 1972). Conversely, newborns and 1-month-olds do not discriminate VOT absolutely categorically (Eimas *et al.*, 1971; Molfese & Molfese, 1979). Two-month-olds do not perceptually integrate two types of fricative acoustic information as adults do (Jusczyk *et al.*, 1979) nor do they discriminate phoneme contrasts under demands on short-term memory (Best *et al.*, 1982; Morse, 1978). These data hint at a perceptual change sometime between 1 and 2½ months, which is consistent with widespread biobehavioral and social changes around 6–10 weeks (e.g., Clifton, Morrongiello, Kulig, & Dowd, 1981; Emde & Robinson, 1979; Haith, 1979), which includes vocal behavior (Oller, 1980; Stark, 1980), suggesting that early changes in speech perception should be further explored (see Werker, 1983, for an example of important age changes in speech perception by older infants).

When infant and adult categorical discrimination has been directly compared, 3-month-olds' (Eimas & Miller, 1980b) and even 7- to 8-month-olds' performance differs significantly from that of adults (Aslin, Pisoni, Hennessy, & Percy, 1981; Eilers, Wilson, & Moore, 1976). Differences from adults, in fact, persist until at least 5–6 years of age (L. E. Bernstein, 1979; Garnica, 1973; Robson, Morrongiello, Best & Clifton, 1982; Schvachkin, 1973; Simon & Fourcin, 1978; Werker & Tees, 1981; Zlatin & Koenigsknecht, 1976). Therefore, boundary effects cannot be considered innate in some absolute sense.

Second, perceptual constancy for vowels is only certain as early as a mean of 2½ months (Kuhl & Miller, 1975). Perceptual constancy for consonants has not been reported earlier than 4 months (Fodor *et al.*, 1975) or 6 months (Holmberg *et al.*, 1977; Kuhl, 1980). Thus, arguments for the innateness of perceptual constancy effects (e.g., Jusczyk, 1981b) should also be held in check.

The exact timing, causes, and nature (i.e., phonetic vs. psychoacoustic) of changes in speech perception cannot be inferred from this literature, however. Even if age had been systematically studied, conclusions would still be limited by the near-exclusive reliance on discrimination measures. The coincidence of an adult phoneme boundary and a peak in infant discrimination is not sufficient evidence to claim "adultlike" perception of the contrast. As argued earlier, assessment of categorical perception requires both labeling and discrimination tests. Simple discrimination cannot reveal whether the distinction was perceived *as a phoneme contrast* (see also Jenkins, 1980), a concern that is equally relevant to the animal research. Discrimination indicates only that *some* difference was detected. Since phoneme discrimination is dissociable from phoneme category identification in aphasics (Blumstein,

Cooper, Zurif, & Caramazza, 1977; Riedel, 1981), it is equivocally involved in aspects of perception that are closer to the meaning of language than mere acoustic contrast detection. Because language-dependent perceptual qualities are more likely to change developmentally, discrimination is inadequate as the sole measure of development in speech perception.

It is uncertain which, if any, speech perception effects are innate, and how they might change developmentally prior to the discovery of words. More crucial for the auditory-phonetic controversy, the following questions remain unanswered: Even if perception of a phoneme contrast is innate, is that innately possessed quality phonetic or psychoacoustic in nature? And if perceptual change does occur during prelinguistic infancy, what is the nature of the change? Each side of the controversy has provided answers (see Aslin & Pisoni, 1980; Trehub, Bull, Schneider, 1981; Walley, Pisoni, & Aslin, 1982; cf. Eilers, 1980; Eilers, Gavin, & Wilson, 1979; Eimas, 1975a), and the auditory-phonetic choice remains unclear. One potential motive force for early perceptual changes can be eliminated, however: if they do occur, they could not have a linguistic motivation from the infant's perspective. This fact causes difficulty for both sides of the controversy, as the next section indicates.

The data on categorical perception of nonspeech, animal perception of phoneme contrasts, perceptual constancy for phonemes, and innateness of infant phoneme perception have thus failed to decide between the psychoacoustic and the phonetic explanations of infant speech perception. When an impasse as extensive as this is reached, it is important to consider whether the difficulty is not with the research but rather with the logic of the two theoretical views themselves.

3. QUESTIONING THE AUDITORY-PHONETIC QUESTION

As stated earlier, the tacit assumptions shared by the two sides of the dichotomy have generally been (a) that the units of speech perception are phonemes (cf., Bertoncini & Mehler, 1981; Jusczyk, 1981a); (b) that some intraperceiver interpretive process must transform acoustic properties to phonemic percepts; and (c) that the process is mediated by some specialized neural mechanism(s). All three assumptions can be questioned, particularly in relation to the infant's discovery of words. The first assumption requires that infants segment phonemes from connected speech. Phonemic segmentation has not been assessed in infants, and is not straightforward even when the perceiver *is* a language-user, since young children seem unable to explicitly segment phonemes (E. J. Gibson & Levin, 1975), as do illiterate adults (Morais, Cary, Alegria, & Bertelson, 1979). The notion that infants perceive phonemes can also be questioned on a linguistic level (see discussion in the next paragraph). The second assumption entails that the listener shed meaning on the presumed meaninglessness of the superficial acoustics of speech, that is, the meaning of the stimulus resides solely in the listener and not directly in the signal. According to the third assumption, the transformation is accomplished by specialized nervous system structures or information-processing stages. Thus, the psychoacoustic and the phonetic views regard speech perception as a mechanistic intraperceiver process.

We turn now to a more detailed examination of each position in the psychoacoustic-phonetic controversy. The main problem with the phonetic view of infant speech perception is that infants presumably do not have a language system, having not yet discovered words. Phoneme contrasts cannot be perceptually available to infants, since they are defined by a language system, being dependent on word meanings in that system. They represent abstract relations among speech sounds that are used by the language to convey semantic differences, as in "pat" versus "bat." Although infants, like adults, categorically distinguish between /p/ and /b/ (e.g., Eimas *et al.*, 1971), they do not necessarily perceive such differences as *phonemic* contrasts (recall that discrimination is an equivocal measure of *phoneme* perception).

The phonetic view also has difficulty accounting for how and why infants would adjust their perceptual categories to suit the language of their environment. Presumably, the evolutionary advantage of innate phoneme categories is that they would filter out irrelevant within-category acoustic variations and thereby relieve the perceiver of having to deal with those unnecessary details. Yet young infants fail to discriminate some phoneme contrasts existing in certain languages according to the adult categories. How and why would infants learning those languages later become able to focus on those innately filtered-out details, in order to adjust their category boundaries or develop new categories? According to researchers on genetic evolution (Jacob, 1977), on nervous system function (Rose, 1976), on perceptual development (Spelke, 1979; Trevarthen, 1979), and on speech perception (Studdert-Kennedy, 1981b), the most efficient evolutionary solution for developmental adaptation to stimulus environments, such as that provided by the native language, is *not* an array of innate mechanisms that are tightly tuned to specific stimulus values, but instead a more flexible attunement to detect the range of stimulus values that could occur. Thus, any specialization we have for perceiving speech would have to be sufficiently flexible to adapt to the specific phoneme contrasts of one's own particular language.

A major drawback of the psychoacoustic view is the argument that the object of speech perception is intrinsically meaningless speech-neutral acoustic data. In particular, this claim has difficulty accounting for the perceived constancy of phonemes spoken by different vocal tracts (vocal-tract normalization) and spoken in different contexts of surrounding phonemes (acoustic variability). The ability to recognize the invariance of a phoneme or word spoken in different contexts and by different people is crucial for language-learning infants. The phonemes they hear occur in a variety of phonemic contexts; and the vocal tracts of the older speakers they hear (and must eventually base their own vocalizations on) differ proportionally from each other, as well as from the infant's own vocal tract. Vocal-tract normalization and acoustic variability do not appear to cause perceptual difficulties for infants. Both infants and other animals apparently solve the normalization and acoustic variability problems in their discriminations among phonemes (e.g., Baru, 1975; Kuhl, 1979, 1980, 1981b; Lieberman, 1980). The psychoacoustic view does not adequately explain the infant's perceptual solution to those problems. The nontrivial acoustic variations involved have thus far defied speaker-independent and speech-neutral acoustic definitions for either phonemes or words. Thus, it cannot be assumed that the infant's solution focuses on speech-neutral acoustic information.

A problematic implication of the psychoacoustic view is that the infant must at some time move from perceiving speech in purely auditory terms to perceiving linguistic structures, such as phonemes (Jusczyk, 1981a, 1981b). What would lead the infant to take the cognitive step from meaningless acoustics to meaning at the level of either phonemes or words? One proposal would be the empiricist philosophical perspective that infants learn meanings by contextual association. However, the associationist solution is unsatisfactory on logical grounds (e.g., E. J. Gibson, 1977; J. J. Gibson, 1966; J. J. Gibson & Gibson, 1955; Jenkins, 1974). Meaning cannot emerge from meaninglessness, so some meaningful element would have to predate the infant's first association.[1] The traditional empiricist perspective is that the elements of meaning are extrinsic to the individual, introduced by sensory stimulation. However, the psychoacoustic view assumes that meaning is *not* intrinsic to the speech stimulus, and it has provided no argument that other stimulation is intrinsically meaningful. In other words, the psychoacoustic view gives us no reason to believe that sensory stimulation in any modality provides extrinsic elements of meaning.

If meaning cannot be assumed to derive from extrinsic stimulation, the traditional nativist perspective offers the alternative proposal that elements of meaning are innate or intrinsic to the infant. The prime candidates for innate elements of meaning in infant speech perception, of course, would be phonemes or phoneme contrasts. This is exactly the position of the phonetic

[1]The general underlying issue of meaning is complex and certainly cannot be settled here. The source of meaning for our knowledge of the world is at the heart of a centuries-long debate in epistemology between phenomenologists and realists. Psychology has taken up this debate. No satisfactory solution, acceptable to all, has been reached in either field. In the context of this chapter, the discussion reflects the author's view on the relation of the topic to how infants perceive speech.

viewpoint questioned previously, and is antithetical to the psychoacoustic viewpoint because it cannot be invoked for animal speech perception. A third possibility is the constructivist perspective that the infant cognitively constructs meaning for sensory stimulation that does not iself provide intrinsic meaning. But this would still depend on some mechanism that determined the nature of the meaning to be constructed, and again the most likely mechanism for speech would be phoneme-based.

A fourth, nontraditional perspective on the source of meaning in speech can be offered, however, which is not consistent with either the psychoacoustic or the phonetic viewpoints. This is the ecological perspective (e.g., J. J. Gibson, 1966) that meaning is directly available to perception in the active, adaptive relation between the perceiver and the objects/events being perceived. It will be discussed at greater length in the subsequent sections of the chapter.

The psychoacoustic position may also be troubled by its evolutionary proposal. It assumes that specialized perceptual "notches" in the sensitivity of the mammalian auditory system along certain phonemically relevant acoustic dimensions have imposed selective pressures on the phonemes that can be uttered by the human vocal tract (e.g., Aslin & Pisoni, 1980; Kuhl, 1981b; J. D. Miller, 1977; Stevens, 1972; Walley et al., 1981). An alternative proposal is that although the anatomy of the human vocal tract places quantal limits on the sounds it can make (see Stevens, 1972); it is this fact that has placed selective pressures on the evolution of specialized notches in the auditory system.

Specializations of neural tissue, like any structural specializations, are naturally selected (evolve) because they have suited some purpose. Yet the psychoacoustic model fails to specify a purpose that could have selected for the speech-related perceptual notches or discontinuities in the mammalian auditory system. Certain basic properties of the auditory system probably are shared by all mammals, reflecting selective adaptation to the commonalities in their auditory environments such as the sounds of weather, vegetation, predators, and prey. However, individual mammalian species do develop more highly specialized sensitivities for certain acoustic properties that are uniquely suited to the particulars of their own ecological niche (e.g., the bat; Neuweiler, Bruns, & Schuller, 1980). For some mammals, notably humans, species-specific vocalizations are particularly important to the species' survival, and have probably placed selective pressures on the development of specialized responsivities of the auditory system. In these species, we would expect to find that specializations in auditory sensitivity have evolved to be uniquely responsive to the vocal characteristics of the species (see also Petersen, 1981; Zoloth, Petersen, Beecher, Green, Marler, Moody, & Stebbins, 1979), which is the converse of the psychoacoustic model of evolution. In the case of speech perception by humans and animals, recall that the specialized notches or discontinuities do indeed appear to be more sensitive and finely tuned in human adults than in the animals studied (see Section 2.4).

The general principle in evolution and in ontogeny of the nervous system has been that motor functions (e.g., vocalization) precede and often motivate the development of the correlated sensory functions (Bekoff, 1981; Horridge, 1968). Consistent with that principle, motor areas develop in advance of sensory areas at each level of the neuraxis (Jacobson, 1978), including the human neocortex (Marshall, 1968; Tuchmann-Duplessis, Aroux, & Haegel, 1975). Even more important, the motor-sensory precedence applies to the neural supports for human speech perception and production: the development and maturation of Broca's area (the motor speech cortex) precedes that for Wernicke's area (the receptive speech cortex) (Rabinowicz, 1979). The more likely evolutionary scenario, then, may be that the human auditory system's properties were selected for best responsiveness to the sound-producing abilities of the uniquely human vocal tract (e.g. Studdert-Kennedy, 1981c), rather than vice versa as the psychoacoustic view suggests.

In summary, both contemporary views of infant speech perception are flawed. The chapter's introduction pinpointed four knotty perceptual problems as requisites to discovering words in the speech stream: (a) disembedding words from connected speech; (b) recognizing word pattern invariances across different utterances and vocal tracts; (c) recognizing the variations that do specify different word patterns; and (d) hearing how to imitate a pattern made by another person. Neither the psychoacoustic nor the phonetic viewpoint offers the infant adequate means for discovering words or phonemes in speech.

But what is the alternative? The remaining sections on *speech source perception* focus on the organization of the speed medium for an answer. The vocal tract offers a structural and dynamic meaning that is intrinsic to speech itself, since as the source of speech it determines the shape of its acoustic product (e.g., Fant, 1973). It would be more parsimonious for perceivers to directly and actively attend to the available vocal-tract information that is intrinsic to speech, than to have the perceptual process mediated with a step involving the meaningful interpretation of meaningless superficial acoustics. According to the *speech source* view, meaning exists in perceiver's relation to speech at its source. This alternative approach derives from the general ecological approach to perception taken by James Gibson and his followers (e.g., J. J. Gibson, 1966; Fowler & Turvey, 1978; Studdert-Kennedy, 1981a, 1981d; Summerfield, 1978; Verbrugge, Rakerd, Fitch, Tuller, & Fowler, in press).

4. AN ECOLOGICAL PERSPECTIVE ON INFANT SPEECH PERCEPTION

Perception of the speech source implies that the infant attends to intrinsically specified information about the vocal tract and the articulatory events that shaped the speech medium. This premise is consistent with the ecological argument that perceiving organisms actively seek information about distal events, which is lawfully specified in the stimulus array (J. J. Gibson, 1966; E. J. Gibson, 1977). It stands in contrast to depictions of perception as the cognitive or neural *transformation* of proximal sensory data, which are intrinsically meaningless and informationally impoverished with respect to the distal event. The speech medium carries many parallel messages, some of which are defined within a particular language, and thus presumably not detected by infants, who do not yet recognize that phonemes can function to distinguish word meanings. Others are human universals, such as the paralinguistic messages of emotional affect, of regional accent, and of age or gender effects on vocal-tract size and configuration.

The speech characteristics of interest for the current discussion are the structural organization and articultory gestures of the vocal tract. According to speech source perception, to perceive these characteristics is to simultaneously apprehend the constant anatomical structure and the transforming positions of articulators in a speaking vocal tract (see also Schubert, 1974). This proposition is supported by the speed and accuracy with which adults and even young children can imitate speech sounds, in spite of the "normalization problem" (e.g., Alekin, Klass, & Chistovich, 1962; Ferguson & Farwell, 1975; Galunov & Chistovich, 1966; Kent & Forner, 1979; see Studdert-Kennedy, 1981a). However, the proposition may seem counterintuitive in two manners. First, common sense suggests that we can *see* the structure and movements of objects, but that we *hear* only "sounds." Thus, the claim that we hear structure and movements, especially the small hidden ones of vocal tracts, may seem unlikely. Second, it may seem implausible that structure and motion are captured at once in the same information, because the qualities of form and movement seem dissociable in our experience. However, an ecological appreciation of sound and hearing shows these "problems" to be false.

4.1. Ecological Acoustics and Speech

Acoustic energy is the radiation over time and space of a wave of rapid alternations in air pressure. It originates from the oscillatory motion of some object or surface that compresses and rarefies the distribution of air molecules around it. Both an object/surface and some oscillatory motion are necessary to produce acoustic energy. In fact, their contributions to sound cannot be dissociated. Structural properties constrain the sound-producing motions an object/surface can undertake; in turn those motions temporally deform the structure in a characteristic manner. It follows that the specific properties of an acoustic flow (e.g., time-varying frequencies and amplitude of the pressure wave) necessarily reflect those structural properties of the object and its vibratory movements that shaped the sound-production.

By the ecological view, the interdependence between structure and transformation is at the core of real events and therefore of their perception (Shaw & Pittenger, 1977). The nature of an object is revealed to the perceiver through event-determined, codefined information about *structural invariants* and *transformational invariants* in objects/events. These terms refer, respectively, to the structural identity of an object undergoing some transformation or change (e.g., by its movement or structural deformation, or by changes in the observer's orientation to it), and the transformations or changes it partakes in. Consider, for example, one person saying three different words as opposed to three people saying a single word. In the first instance, information common to the three words reflects the structural invariance of that speaker's vocal tract. In the second case, there is a transformational invariant in the articulation of the single word by three structurally different vocal tracts. Structural and transformational invariants lawfully shape the energy medium that carries their message (i.e., acoustical or optical energy). The ecological premise is that, through their modulation of the energy medium, transformations can perceptually specify an object's structure. In support of this, infants' visual recognition of objects and their structure is enhanced by watching the objects undergo various spatial-temporal transformations (E. J. Gibson, 1980; Ruff, 1980, 1982).

From the ecological perspective, auditory perception is the codetection of the transformations (motions) and structure of the sound source, which are veridically conveyed in the acoustic medium (see also Schubert, 1974; Warren & Verbrugge, in press). Thus, structure and motion are not only seen but heard. In the case of speech, the acoustic medium is better suited than optics for conveying the structure and transformations of the vocal tract, many of which are invisible in face-to-face communication as well as when the speaker is out of view. Articulatory gestures may be beyond the capabilities of vision in another way, since their speed and precision exceed the temporal and spatial resolution of the visually perceived manual American Sign Language (Studdert-Kennedy & Lane, 1980).

The unseen messages carried by natural speech reflect not only the origin of its acoustic energy (respiratory and laryngeal), but also the structural identity, biokinematic coupling, and specific movements of the speaker's supralaryngeal vocal tract.[2] In vocalizations and musical sounds (among others), the acoustic wave does not radiate freely from its oscillatory origin to the perceiver: the medium is also molded by the structure and transformations of an intervening resonating tube. The size and shape of a resonant cavity determine its natural resonating frequency (or frequencies) at which the air contained within its walls will oscillate when excited by a flow of air introduced from outside. If the extrinsic air flow is already oscillating (e.g., when acoustic energy from the vibrating larynx is introduced to the supralaryngeal vocal tract), then those oscillatory frequencies in the flow that match the resonant properties of the tube will be amplified in intensity; other frequencies that mismatch the resonant properties will be attenuated (filtered out). The larger a resonating cavity is, the lower its primary resonant frequency, which is also affected by the size and number and positions of openings in the cavity. As its shape deviates from perfectly spherical or cylindrical, particularly if there are corners or "side pockets," higher order resonant frequencies may be added. Surface properties, such as the smoothness and elasticity of the resonant tube's walls, largely determine the time course of intensity changes in the resonated frequencies.

In the case of speech, the critical sound-shaping properties of the resonant tube include the shape and elasticity of the cheeks, throat, lips, and tongue, and the position and rigidity of the teeth. The properties that allow the vocal tube to transform in shape are especially important for its articulatory gestures: the hinged movements of the jaw, and the moving and deforming obstructions of the tongue, lips, and velum (which opens the nasal passage to resonate for sounds like /m/ and /n/). These structural and transformational properties all shape the acoustic speech wave. The time-varying formants and other acoustic discontinuities (see Section 1.3) reflect, in particular, rapid transformations of the vocal-tract resonant configuration that are produced by movements of the tongue, lips, jaw, and velum, within the constraints posed by the enduring anatomical relations within the tract. (For more detailed discussions of the physical acoustics of speech, see Fant, 1973; Flanagan, 1973.)

Since these distal source properties determine the acoustic shape of speech, they should be available to perceivers (see also Studdert-Kennedy, 1981a, 1981d; Summerfield, 1978; Verbrugge et al., in press). Infants and even animals should be able to detect at least some of the structural and

---

[2]Natural speech also identifies the speaker as a member of the human species. Synthetic speech, insofar as it "works" perceptually, must capture necessary information about human vocal-tract dynamics, and usually also about the structure of a generic vocal tract, often appropriate for an adult male of indeterminate age. Rarely, however, does synthetic speech capture sufficient "textural" detail about a natural vocal tract to sound like a live human speaker, even an unknown one.

transformational invariants in speech, although evolutionary and ontogenetic history will affect how well different perceivers are attuned to pick up the various messages. As discussed earlier, although animals and human infants do show phoneme boundary effects, their performance deviates significantly from human adults' categorical perception.

The speech source perception view may be further clarified by comparison and contrast with the psychoacoustic and the phonetic views. If auditory perception is the detection of information about source structure and transformations in the acoustic medium, then the perception of speech should be abstractly similar to the perception of other sounds, in agreement with the psychoacoustic view. However, the speech source perspective disagrees with the psychoacoustic notions that the perceiver's focus is on event- or speech-neutral acoustic parameters, and that these parameters need to be transformed by the system into percepts. The speech source proposal is also consistent in an important respect with the phonetic view. That is, speech perception even by the infant is considered "special" and uniquely human; however, that special perceptual quality is not agreed to be based on phonemes for infants. Moreover, the relative emphasis on "speech" and "perception" is different. Because the speech medium conveys its source to the perceiver, acoustic cues need not be transformed into percepts, whether by codes for phoneme categories or by neuromotor codes for phoneme production.

According to the speech source view, the specialness of speech perception derives from the unique structural and transformational properties of its sound source, the human vocal tract. It is unique in its complex anatomy, its biokinematic organization, and its particular dynamic gestures (e.g., Lieberman, 1967; Lieberman et al., 1971), all of which are reflected in its acoustic productions. Moreover, humans have a privileged relation to speech as the tool of human-specific language communication.

Some of the advantages of this view for several important aspects of speech perception in adults and infants will be considered next.

4.2. Speech Source Perception in Adults

The major acoustic puzzles in speech perception research have been the problems of acoustic variability and normalization (see Section 1.3.). As a reminder, the acoustic variability problem refers to the sometimes quite striking variability in the acoustic properties of an invariantly perceived phoneme, which occurs primarily when it is produced in different contexts of surrounding phonemes. The variability is caused by coarticulation among phonemes as well as by the articulatory trajectories that interconnect adjacent phonemes. The acoustic properties of the phoneme are thus assimilated to the acoustic properties of its neighbors (e.g., in Figure 1, the differences among the /l/'s in "little" and "girl"). Another source of acoustic variability is the wide variety of acoustic features that can identify a given phoneme (e.g., Lisker, 1978). These sorts of acoustic variations pose a greater empirical puzzle for psychoacoustic accounts of consonant perception rather than vowel perception. They have much stronger effects on the formant trajectories and other acoustic features associated with consonants than on the formant frequencies in the nuclear portions of vowels (although the latter are also affected to considerable degree in conversational speech).

The normalization problem refers to another acoustic context effect, caused by variations in the dimensions of different vocal tracts. It causes greater difficulties for psychoacoustic explanations of vowel perception than consonant perception. The formant frequencies in vowel nuclei are more obviously affected by vocal-tract proportions than are the formant trajectory patterns associated with consonants. The different effects of these two types of acoustic variability in vowels and consonants suggest a difference in the information those two phoneme classes convey, which has been supported by several other lines of research on speech production (e.g., Fowler, 1980; Fowler et al., 1980) and perception (e.g., Ades, 1977; Crowder, 1973; Cutting, 1974; Darwin, 1971; Pisoni, 1973; Studdert-Kennedy & Shankweiler, 1970).

That adults perceive an invariant identity underlying the acoustical variations of a given consonant (e.g., acoustic differences for /d/ in "dee" versus "dah" versus "doo") is difficult for the psychoacoustic view to explain, because it requires finding a unitary acoustic principle for the invariant percept. Although several auditory solutions have been proposed for the acoustic invariance problem, for example, perceptual "templates" for consonant-specific spectral (frequency) acoustic properties (Blumstein, 1980; Searle, Jacobson, & Rayment, 1979; Stevens & Blumstein, 1978), these do not hold well up under empirical test (Blumstein, Isaacs, & Mertus, 1982; Walley et al., 1981) or logical scrutiny (e.g., Liberman, 1982; Studdert-Kennedy, 1981a,d). However, the acoustic variability is actually an advantage from the speech source perception view that speech acoustics convey information about the structural and transformational invariants of their vocal-tract source. That is, for the variability that derives from the coarticulation of adjacent phonemes, the form of that vocal-tract transformation should clarify rather than confuse the source properties that identify both elements. As for the variety of acoustic features that can specify a given consonant, these also result from, and thus may offer equivalent information about, the vocal-tract invariants identifying that consonant.

4.2.1. Consonant Perception

Research on phoneme context effects has found shifts in category boundary positions that are predictable from the coarticulatory effects of different neighboring phonemes. For example, a continuum between "s" and "sh" can be generated by varying only the center frequency of the fricative noise. But the frequency of natural fricatives is lower if the following vowel is "oo" rather than "ah." That is because the lip-rounding for "oo," which lengthens the vocal tract and therefore lowers its resonant frequencies, is coarticulated with the fricative (e.g., Bell-Berti & Harris, 1979). In support of the notion that perceivers detect coarticulatory information, the "soo-shoo" boundary occurs at a lower frequency than the "sah-shah" boundary (Mann & Repp, 1980). Similar coarticulatory context effects have been found for stop consonant place of articulation differences (Mann, 1980) and for stop consonant voicing boundaries (e.g., Summerfield, 1982).

Research on the perceptual unity of multiple acoustic properties for a consonant distinction offers converging support for the speech source position. The various acoustic properties are not perceived according to their acoustic differences but are perceived instead as equivalent information about the articulation of the same consonant (Best, Morrongiello, & Robson, 1981; Fitch, Halwes, Erickson, & Liberman, 1980). These perceptual equivalences and context effects are difficult to explain by speech-neutral psychoacoustic mechanisms (see Studdert-Kennedy, 1981d), and control studies have in fact failed to find analogous effects in nonspeech perception (Best et al., 1981; Mann, Madden, Russell, & Liberman, 1981; Summerfield, 1982).

4.2.2. Vowel Perception

If vowel perception is accomplished by detecting the underlying vocal-tract configuration and transformations that remain invariant across the structural variations of different vocal tracts, then vocal-tract normalization would not be a problem for speech source perception. In contrast, the psychoacoustic account posits that a singular underlying neutral acoustic description of the vowel must be derivable by some formula. No such description has yet been found because proportional differences among vocal tracts, especially male versus female versus child, prevent a uniform scaling of vowel formant frequencies among speakers (Broad, 1981); that is, formant frequency ratios are speaker-specific. Nor can the problem be solved through some formula that partials out sex and age differences based on the value of some independent acoustic feature such as fundamental frequency of the voice, which is the most obvious formant-independent feature that could differentiate those speaker characteristics. The sex and age groups show considerable overlap in fundamental frequency, a laryngeal property that is imperfectly correlated with the variation in supralaryngeal configurations that affect formant properties. Furthermore, as the latter observation suggests, the perceived sex and age of a speaker depend on supralaryngeal characteristics rather than on fundamental frequency (Lehiste & Meltzer, 1973). Thus, the psychoacoustic approach is left in an untenable position: the acoustic normalization solution would appear to depend on *a priori* knowledge of the supralaryngeal vocal-tract properties whose influence it is trying to circumvent.

Of further relevance to the speech source view, identification performance is better for vowels spoken in CVC context than for isolated vowels, even though formant frequencies are more clearly differentiated among the isolated vowels (Strange, Verbrugge, Shankweiler, & Edman, 1976). Likewise, similarity judgments among vowels in CVCs are clearly differentiated along three dimensions, which correspond closely to vowel articulatory factors, whereas most perceivers differentiate isolated vowels along only one or two dimensions (Rakerd & Verbrugge, 1982). These contextual effects suggest that coarticulatory information aids vowel as well as consonant perception, consistent with the ecological premise that structural invariants (vocal tract configurations) are clarified by transformational (dynamic articulation) properties (see Verbrugge, Shankweiler, & Fowler, 1980). Studies with CVC syllables whose vowel nucleus has been replaced with silence, leaving only the syllable-initial and syllable-final formant transitions (in correct temporal relation), further support the ecological interpretation. Vowel identification under these conditions is remarkably well-preserved (Strange, Jenkins, & Edman, 1977), even if each remaining piece of coarticulatory information is taken from a different-sexed speaker (Verbrugge & Rakerd, 1980). These coarticulatory influences on vowel perception are no problem for the view that we detect vocal-tract source information, but would be difficult to explain via psychoacoustic mechanisms (Fowler & Shankweiler, 1978; Shankweiler, Strange, & Verbrugge, 1977; but see Howell, 1981).

These consonant and vowel findings suit the speech source interpretation, but at the same time they also suit a phonetic interpretation of *adult* speech perception. Certainly, the experimental tasks often required subjects to "recover phonemes from the speech stream" (Liberman, 1982). Since speech conveys language, the detection of "pure" (nonlinguistic) structural and transformational invariants of the vocal tract may rarely if ever be ends in themselves for adults, and instead serve as means to the *linguistic* ends of recognizing, for example, phonemes. This would not be the case for infants, however, since phonemes can be "recovered" from the speech stream only by listeners who already know they are there (Studdert-Kennedy, 1981d). Phonemes may indeed be at the "surface of language" (Liberman, 1982) for adults, but vocal-tract source information must be at the "surface of speech" for prelinguistic infants. We now turn to some recent research with prelinguistic infants, which offers strong support for speech source perception.

4.3. Speech Source Perception by Infants

4.3.1. Infants' Perceptual Constancy for Speech

Perceptual constancy is at the base of the acoustic variability and vocal-tract normalization puzzles. The findings just discussed suggest that the adult's solution lies in a perceptual focus on the structural and transformational invariants of a speaker's vocal tract. Even without the adult's linguistic motivations, infants also show perceptual constancy for vowels and consonants spoken by different people or in different phoneme contexts, as described earlier in the chapter (Holmberg et al., 1977; Kuhl, 1979, 1980, 1981b). Therefore, the invariant features they apprehend must exist at the surface of speech and not only at the surface of language. The speech source view suggests that the properties of relevance to the infant lie in the vocal tract and not in the speech-neutral superficial acoustics.

A psychoacoustic interpretation of infant perceptual constancy works no better than it did for adults. We cannot assume that infants are guided by knowledge about phonemes in their solution of the two acoustic puzzles, so some independent source of guidance to the invariant acoustic features of vowels and consonants would be needed. One psychoacoustic solution to the normalization problem might be that although adults do not solve it by partialing out speaker differences based on fundamental frequency, linguistically naive infants do. However, this approach does not work, because infants as well as adults appear to rely on supralaryngeal information in the formant structure of speech rather than on fundamental frequency when perceiving speaker gender (C. L. Miller, Younger, & Morse, 1982). Nor does an acoustic template model (e.g., Blumstein) appear to give an adequate psychoacoustic explanation to the acoustic invariance problem of infants' perceptual constancy for consonants across varying vowel contexts (Studdert-Kennedy, 1981d; Walley et al., 1981).

What *does* remain constant in the utterances of a vowel or consonant by different speakers or across different phoneme contexts is the underlying similarity in vocal-tract structure and articulator positioning. Speech source information would thus seem to offer a more straightforward metric than speech-neutral acoustic invariance for infant perceptual constancy, as was argued in the case of adult speech perception. Moreover, perception of speech source information would certainly be a more direct guide than speech-neutral acoustic patterns for the infant's attempts at vocal imitation of older speakers and eventual production of words provided by her native language environment.

Thus far, the central argument that the infant perceives speech source information has been oriented around the acoustic medium. However, as indicated in the next section on infants' recognition of auditory and visual commonalities in speech, this information is provided by sight as well as by sound. The intermodal perception of speech by infants provides strong support for the speech source perception view, and is particularly difficult to reconcile with the psychoacoustic and phonetic perspectives.

4.3.2. Infants' Intermodal Perception of Speech

When adults listen and watch someone speak, the acoustic and optic information about speech is not perceptually independent. Rather, the two seemingly disparate types of information are perceptually unified, implying that a common metric underlies them. Prelinguistic infants likewise recognize underlying commonalities between audio and visual presentations of speech that cannot be described in linguistic terms for them. The speech source perception view suggests that infants perceive speech intermodally by attending to the underlying articulatory events that provide the auditory and visual information.

To appreciate the contribution of visual information to speech perception, recall listening to someone speaking at the front of a room. It probably seemed easier to understand what was being said if you could also keep the speaker's face in view. This intuition has recently been empirically validated with adult listeners. Under difficult listening conditions, adults perceive speech more correctly when they can watch the speaker than when they must rely on their ears alone (Binnie, Montgomery, & Jackson, 1974; Dodd, 1977; Summerfield, 1979). These findings suggest that listeners obtain information about speech not only from the acoustic signal, but also from the optical information that results from articulatory maneuvers.

Of course, the major responsibility for speech perception is carried by the auditory modality. That blind adults successfully perceive speech, whereas the deaf have serious difficulty with lipreading, would seem to imply that auditory information is both necessary and sufficient for speech perception, although visual information plays a negligible role unless listening is particularly difficult. Speech researchers accepted this logic until recently, when MacDonald and McGurk (1978) and Summerfield (1979) reported that listeners fail to recognize phonemic conflicts in concurrent auditory and visual presentations of speech, instead perceiving a unified speech event. The percepts did not veridically reflect either the acoustic or the optic signal considered in isolation. For example, when perceivers watched a face silently articulating "ga" while a voice said "ba," they heard "da." These results indicate that in face-to-face speech perception, listening is not simply supplemented by arbitrary, learned associations between vocal-tract configurations and speech sounds. Rather, at the level of the speech event itself, the information provided by the two modalities shows an intermodal articulatory equivalence.

Two opposing views have been offered for adults' perception of acoustic-optic equivalence in speech. MacDonald and McGurk (1978) have suggested that the equivalence be described linguistically, in terms of abstract features of phonemes. The other view, proposed by Summerfield (1979) and based on the Fowler *et al.* (1980) ecological interpretation of speech production findings, argues that the equivalence is nonlinguistic and modality-free, arising from the dynamics of articulation. The first view has been criticized, in part, because it accounts for only a limited number of the speech percepts that result from audiovisual conflict (see Summerfield, 1979, for detailed discussion of these and other criticisms).

In light of recent findings, the latter view appears to best account for how prelinguistic infants perceive speech intermodally. Infants' sensitivity to acoustic-optic equivalences in speech have been demonstrated under two conditions. Under the first condition, infants were presented with acoustic-optic speech displays in which the overall synchrony and the specific articulatory details presented in the two modalities of information were confounded. In one study, 2½- to 4-month-old infants saw the mirrored reflection of a woman's face repeating nursery rhymes; the auditory signal was either in synchrony or delayed relative to the optic presentation by 400 milliseconds. The infants watched the reflection significantly longer when the visual and the auditory presentations were in natural synchrony (Dodd, 1979). In a second study, 4-month-old infants viewed two women speaking, in two adjacent video films, while the concurrent speech of one woman was played over a central loudspeaker. Infants preferred to look at the face that talked in synchrony to the audio speech presentation (Spelke & Cortelyou, 1981).

It is unclear whether the infants were responding to the general synchrony and/or the specific articulatory details of the optic and acoustic displays, however, since these two aspects of audiovisual match were confounded in both experiments. They might have only recognized the overall synchrony, for example, for syllable onsets, between the speech seen and heard. However, they might also have preferred watching the natural acoustic-optic concurrence of specific articulatory gestures. For example, infants might prefer to look at a speaker's lips being rounded and protruded for the production of the vowel "oo" as they heard an audio "oo," as opposed to looking at the speaker's lips being opened wider to produce "ah."

This prediction was recently tested experimentally (Kuhl & Meltzoff, 1982; MacKain, Studdert-Kennedy, Spieker, & Stern, 1981). Kuhl & Meltzoff (1982) presented 4 to 5-month-olds with two adjacent films of a woman's face synchronously articulating the vowels "ah" and "ee," while one of those vowels was presented auditorily and in synchrony over a central loudspeaker.

The infants preferred to watch the film whose articulatory details specified the vowel presented auditorily. In the MacKain *et al.* study, disyllables (e.g., "mama," "lulu") were presented audiovisually, under similar experimental conditions, to 5- to 6-month-old infants. The infants looked significantly longer at the video display whose articulatory dynamics matched the acoustic presentations, for the disyllables "mama," "baby," and "zuzu." These findings indicate that young infants recognize at least some auditory-visual equivalences of articulatory gestures, and are not only sensitive to general synchrony. Moreover, this intermodal recognition was accomplished in the presumed absence of a language system, making linguistically based explanations (e.g., MacDonald & McGurk) untenable for this age group. The commonality that the infants recognized between the acoustic and the optic information may best be described in nonlinguistic terms (Summerfield, 1978).

Given that the infant seems attuned to detect vocal-tract source information in speech, what may be the organization of the supporting perceptual system? A consideration of the biological basis of this attunement should move us closer to understanding the ease with which humans recognize auditory-visual equivalences in articulatory details, and the infant's apparent ease in learning to speak a first language. Research with adults suggests that the answer lies in the functional asymmetries of the left- and right-cerebral hemispheres of the human brain.

4.4. Left-Hemisphere Attunement for Articulatory Information

For the adult, the left-cerebral hemisphere shows a specialized advantage for the perception of speech, in contrast to a right-hemisphere advantage for the perception of music and certain other nonspeech sounds (e.g., Kimura, 1973). Even in young infants, the hemispheres are differentially responsive to human speech versus other sounds. Auditory evoked response asymmetries in young infants favor the left hemisphere when words or syllables are presented auditorily, whereas the right-hemisphere response is stronger when musical or other nonspeech sounds are presented (Molfese, Freeman & Palermo, 1975). In dichotic listening tests, consistent with the adult findings previously cited, infants as young as 2½ to 3 months show a right-ear advantage (REA) in discriminating among consonants, indicating a left-hemisphere superiority. Conversely, they show a left-ear advantage (LEA) in discriminating notes played by different musical instruments, indicating a right-hemisphere superiority (Best, Hoffman, & Glanville, 1982; Entus, 1977; Glanville, Best, & Levenson, 1977).

These functional asymmetries in infants indicate an early left-hemisphere attunement to information in speech, which could be an important biological support for the infant's perceptual discovery of the articulatory patterns of spoken words. But the data do not indicate exactly the sort of information in speech to which the left hemisphere is attuned (see Molfese, Nuñez, Seibert, & Ramaniah, 1976). Two recent findings suggest that the infant's left hemisphere is apparently attuned to information about the articulatory gestures of the vocal tract.

As an additional result of their intermodal speech perception study, MacKain, Studdert-Kennedy, Spieker, & Stern (1983) found a rightward attentional bias (implying left-hemisphere activation: Kinsbourne, 1973, 1982), which facilitated the infants' recognition of acoustic-optic commonality in the articulatory details of speech. In that experiment, infants attended primarily to either the right or the left video monitor during the synchronous audio presentation. An analysis of visual preferences indicated that the infants recognized auditory-visual matches versus mismatches in articulatory properties only when they were attending to the *right* video monitor. Since intermodal perception of speech appears to entail the recognition of its vocal-tract source properties, these results indicate the infant's recognition of that information is facilitated by a left-hemisphere attentional bias toward those properties.

The results of another study (Best, 1978) may further clarify which aspects of human speech are the object of the infant's left-hemisphere attunement. Adults show a consistent left-hemisphere advantage for consonant perception, whereas isolated vowels yield a nonsignificant perceptual asymmetry (e.g., Studdert-Kennedy & Shankweiler, 1970). Even vowels in CVC syllables show a weak and equivocal left-hemisphere advantage (e.g., Studdert-Kennedy & Shankweiler, 1970; Weiss & House, 1973). This vowel-consonant difference in hemispheric perceptual asymmetry may depend on the earlier discussed differences in the acoustic and articulatory properties of vowels and consonants.

The aim of the infant study (Best, 1978) was to determine whether 3½-month-olds show a similar hemispheric difference in perception of consonants versus vowels. The infants showed a clear left-hemisphere advantage for discriminating a set of synthetic consonants, as adults did. However, the infants also showed a clear *right*-hemisphere advantage for discriminating steady-state synthetic vowels, which differs from adult reports.

The psychoacoustic interpretation of the adult hemisphere differences is that the hemispheres are differentially specialized for processing the acoustic features that differ between consonants and vowels (e.g., Cutting, 1974; Schwartz & Tallal, 1980). However, this interpretation confounds acoustic and articulatory differences between vowels and consonants, and must be rejected on methodological grounds (Studdert-Kennedy & Shankweiler, 1980), as well as for the general criticisms against the psychoacoustic view. A speech source interpretation would instead consider articulatory differences between consonants and vowels. Indeed, consonant and vowel productions engage different coordinations of the articulatory musculature (Fowler, 1980). Moreover, vowel-consonant differences in intermodal speech perception effects (Summerfield, 1979; Summerfield, McGrath, & Forster, 1982) suggest that such articulatory differences may be influential in perception. The pattern of the phoneme class differences in production and in intermodal perception suggest that consonant information is conveyed in rapid articulatory changes, whereas vowel information is conveyed in relatively more slowly changing configurations of the tongue, lips, and jaw.

The implication of these two recent findings on infant hemispheric asymmetries is that the left-hemisphere attunement takes the form of an attentional bias toward information about rapid articulatory transformations. In complement to that attentional bias, the infant's right hemisphere may be better attuned to information about relatively more enduring structural properties of sound-making objects, such as the structural properties of instruments that determine their musical timbre and the configurations of the articulators in the human vocal tract that determine steady-state vowel color.

In conclusion, these specializations of the cerebral hemispheres for responding to different aspects of articulatory information in speech may offer biological support for the infant's discovery and production of words. In the final section, the relation of speech source perception to the discovery of words and the broader motivation provided by the context of communicative development will be briefly discussed.

5. THE BROADER CONTEXT OF COMMUNICATIVE DEVELOPMENT

The speech medium carries a number of parallel messages (Pike, 1959), and not only the sort of vocal-tract source information we have been focusing on at the surface of speech. To learn language, the infant must discover, or learn to recognize, many or all of those other messages as they are specified by convergent information in speech and in the context of its occurence. Some of the messages that must be discovered are linguistic, beginning with words or phrases. Although the linguistic messages are more abstract than speech source messages, their expression in the speech medium depends directly on speech source information. Words and phrases are conveyed in the medium as patterns of articulator configurations and transformations, which have invariant properties across speakers, speaking rates, and surrounding speech context. Therefore, the infant's eventual discovery of them depends on attention to informational invariants in vocal-tract shapes and gestures as they are patterned over time, and conveyed in both the acoustic and optic media.

As for phonemes, their discovery as invariant vocal patterns that convey differences in meaning appears to be *served by* the child's developing use of words rather than vice versa (Menn, 1980; Menyuk & Menn, 1979). Given that the function of phonemes in a language system is determined by word meanings and contrast, it is consistent with this interpretation that maternal speech to toddlers who produce words *does* include hyperdifferentiation in the productions of some phoneme contrasts, whereas maternal speech to prelinguistic infants does *not* include such hyperdifferentiation (Malsheen, 1980).

But the recognition of invariant vocal patterns, of course, is still not support enough for the discovery of words and phonemes. In order to comprehend and use language normally, the infant must also come to recognize the speaker's *communicative* messages. Some of these appear at the surface of speech, where infants can detect them. Emotional affect may be conveyed to the infant directly in the mother's speech, for example, through the level and modulations of her voice pitch and intensity (see Stern, Spieker, & MacKain, 1982). These communicative messages often gain converging support from information in the visual and even in the haptic modalities; for example, emotional affect often receives contextual support in the speaker's changing facial expressions and the way she or he touches or holds the infant. The argument here is that the infant's discovery of more abstract communicative messages, notably the referential meaning of words, *requires* such nonspeech contextual support. Word meanings cannot be revealed for the first time through the speech medium alone.

This observation brings us to the end of our discussion, moving as it does beyond both the prelinguistic period and the infant's perception of the surface of speech. In closing, however, it is suggested that the prelinguistic infant's ability to perceive in speech the structure and transformations of its vocal-tract source provide her or him a crucial tool for discovering the more abstract messages of language.

Acknowledgements

I would like to thank Steven S. Braddon, Kristine S. MacKain, Roy Pea, Carol Fowler, Michael Studdert-Kennedy, and the editors of this book for their thoughtful comments on an earlier version of this chapter. Special thanks are due to Kristine S. MacKain for her helpful discussion about intermodal speech perception by infants, and to Steven Braddon for discussion and encouragement throughout the project.

6. REFERENCES

Ades, A. E. Vowels, consonants, speech and nonspeech. *Psychological Review*, 1977, *84*, 524–530.

Alekin, R. O., Klaas, Y. A., & Chistovich, L. A. Human reaction time in the copying of aurally perceived vowels. *Soviet Physics-Acoustics*, 1962, *8*, 17–22.

Aslin, R. N., & Pisoni, D. B. Some developmental processes in speech perception. In G. Yeni-Komshian, J. F. Kavanaugh, & C. A. Ferguson (Eds.), *Child phonology. Vol. 2: Perception.* New York: Academic Press, 1980.

Aslin, R. N., Pisoni, D. B., Hennessy, B. L., & Percy, A. J. Discrimination of voice onset time by human infants: New findings and implications for the effects of early experience. *Child Development*, 1981, *52*, 1135–1145.

Baru, A. V. Discrimination of synthesized vowels [a] and [i] with varying parameters in dog. In G. Fant & M. A. A. Tatham (Eds.), *Auditory analysis and the perception of speech.* London: Academic Press, 1975.

Bekoff, A. Embryonic development of the neural circuitry underlying motor coordination. In W. M. Cowan (Ed.), *Studies in developmental neurobiology.* New York: Oxford University Press, 1981.

Bell-Berti, F., & Harris, K. S. Anticipatory coarticulation: Some implications from a study of lip-rounding. *Journal of the Acoustical Society of America*, 1979, *65*, 1268–1270.

Bernstein, J. Formant-based representation of auditory similarity among vowel-like sounds. *Journal of the Acoustical Society of America*, 1981, *69*, 1132–1144.

Bernstein, L. E. Developmental differences in labeling VOT continua with varied fundamental frequency. *Journal of the Acoustical Society of America*, 1979, *65*, S112 (A). (Supplement)

Bertoncini, J., & Mehler, J. Syllables as units in infant speech perception. *Infant Behavior and Development*, 1981, *4*, 247–260.

Best, C. T. *The role of consonant and vowel acoustic features in infant cerebral asymmetries for speech perception.* Unpublished doctoral dissertation, Michigan State University, 1978.

Best, C. T., Morrongiello, B., & Robson, R. Perceptual equivalence of two acoustic cues in speech and nonspeech perception. *Perception and Psychophysics*, 1981, *29*, 191–211.

Best, C. T., Hoffman, H., & Glanville, B. B. Development of infant ear asymmetries for speech and music. *Perception and Psychophysics*, 1982, *31*, 75–85.

Binnie, C. A., Montgomery, A. A., & Jackson, P. L. Auditory and visual contributions to the perception of consonants. *Journal of Speech and Hearing Research*, 1974, *17*, 619–630.

Blumstein, S. E. Speech perception: An overview. In G. H. Yeni-Komshian, J. F. Kavanaugh, & C. A. Ferguson (Eds.), *Child phonology. Vol. 2: Perception.* New York: Academic Press, 1980.

Blumstein, S. E., Cooper, W. E., Zurif, E., & Caramazza, A. The perception and production of voice-onset time in aphasia. *Neuropsychologia*, 1977, *15*, 371–383.

Blumstein, S. E., Isaacs, E., & Mertus, J. The role of the gross spectral shape as a perceptual cue to place of articulation in initial stop consonants. *Journal of the Acoustical Society of America*, 1982, *72*, 43–50.

Broad, D. J. Piecewise planar vowel formant distributions across speakers. *Journal of the Acoustical Society of America*, 1981, *69*, 1423–1429.

Burdick, C. K., & Miller, J. D. Speech perception by the chinchilla: Discrimination of sustained /a/ and /i/. *Journal of the Acoustical Society of America*, 1975, *58*, 415–427.

Clifton, R. K., Morrongiello, B. A., Kulig, J. W., & Dowd, J. M. Developmental changes in auditory localization in infancy. In R. N. Aslin, J. R. Alberts, & M. R. Petersen (Eds.), *Development of perception: Psychobiological perspectives.* (Vol. 1.) New York: Academic Press, 1981.

Cooper, F. S. Some reflections on speech research. *Haskins Laboratories: Status Report on Speech Research,* 1981, *SR-65,* 201–215.

Crowder, R. G. Representation of speech sounds in precategorical acoustic storage. *Journal of Experimental Psychology,* 1973, *1,* 14–24.

Cutting, J. E. Two left hemisphere mechanisms in speech perception. *Perception and Psychophysics,* 1974, *16,* 601–612.

Cutting, J. E., & Eimas, P. D. Phonetic feature analyzers and the processing of speech in infants. In J. F. Kavanaugh & J. E. Cutting (Eds.), *The role of speech in language.* Cambridge: M.I.T. Press, 1975.

Cutting, J. E., & Rosner, B. S. Categories and boundaries in speech and music. *Perception and Psychophysics,* 1974, *16,* 564–570.

Darwin, C. J. Ear differences in the recall of fricatives and vowels. *Quarterly Journal of Experimental Psychology,* 1971, *23,* 46–62.

Dodd, B. The role of vision in the perception of speech. *Perception,* 1977, *6,* 31–40.

Dodd, B. Lip reading in infants: Attention to speech presented in- and out-of-synchrony. *Cognitive Psychology,* 1979, *11,* 478–484.

Dudley, H. The carrier nature of speech. *Bell System Technical Journal,* 1940, *19,* 495–515. (Reprinted in J. L. Flanagan & L. R. Rabiner (Eds.), *Speech synthesis.* Stroudsberg, Penn.: Dowden, Hutchinson & Ross, 1973.)

Eilers, R. E. Infant speech perception: History and mystery. In G. Yeni-Komshian, J. F. Kavanaugh, & C. A. Ferguson (Eds.), *Child phonology. Vol. 2: Perception.* New York: Academic Press, 1980.

Eilers, R. E., & Gavin, W. J. The evaluation of infant speech perception skills: Statistical techniques and theory development. In R. E. Stark (Ed.), *Language behavior in infancy and early childhood.* New York: Elsevier-North Holland, 1981.

Eilers, R. E., & Minifie, F. D. Fricative discrimination in early infancy. *Journal of Speech and Hearing Research,* 1975, *18,* 158–167.

Eilers, R. E., Wilson, W. R., & Moore, J. M. Discrimination of synthetic prevoiced labial stops by infants and adults. *Journal of the Acoustical Society of America,* 1976, *60,* S91 (A). (Supplement)

Eilers, R. E., Wilson, W. R., & Moore, J. M. Developmental changes in speech discrimination in infants. *Journal of Speech and Hearing Research,* 1977, *20,* 766–780.

Eilers, R. E., Oller, D. K., & Gavin, W. J. *A cross-linguistic study of infant speech perception.* Paper presented at the meeting of the Southeastern Conference on Human Development. Atlanta, Georgia, 1978.

Eilers, R. E., Gavin, W., & Wilson, W. R. Linguistic experience and phonemic perception in infancy: A cross-linguistic study. *Child Development,* 1979, *50,* 14–18.

Eimas, P. D. Linguistic processing of speech by young infants. In R. L. Schiefelbusch & L. L. Lloyd (Eds.), *Language perspectives—Acquisition, retardation, and intervention.* Baltimore: University Park Press, 1974. (a)

Eimas, P. D. Auditory and linguistic processing of cues for place of articulation by infants. *Perception and Psychophysics,* 1974, *16,* 513–521. (b)

Eimas, P. D. Speech perception in early infancy. In L. B. Cohen & P. Salapatek (Eds.), *Infant perception: From sensation to cognition. Vol. 2: Perception of space, speech, and sound.* New York: Academic Press, 1975. (a)

Eimas, P. D. Auditory and phonetic coding of the cues for speech: Discrimination of the [r-l] distinction by infants. *Perception and Psychophysics,* 1975, *18,* 341–347. (b)

Eimas, P. D., & Miller, J. L. Contextual effects in infant speech perception. *Science*, 1980, *209*, 1140–1142. (a)

Eimas, P. D., & Miller, J. L. Discrimination of information for manner of articulation. *Infant Behavior and Development*, 1980, *3*, 367–375. (b)

Eimas, P. D., Siqueland, E., Jusczyk, P., & Vigorito, J. Speech perception in infants. *Science*, 1971, *171*, 303–306.

Emde, R. N., & Robinson, J. The first two months: Recent research in developmental psychobiology and the changing view of the newborn. In J. Noshpitz & J. Call (Eds.), *Handbook of child psychiatry: Vol. 1: Development*. New York: Basic Books, 1979.

Entus, A. K. Hemispheric asymmetry in processing of dichotically-presented speech and non-speech stimuli by infants. In S. J. Segalowitz & F. A. Gruber (Eds.), *Language development and neurological theory*. New York: Academic Press, 1977.

Fant, G. *Acoustical theory of speech production*. The Hague: Mouton, 1960.

Fant, G. *Speech sounds and features*, Cambridge: M.I.T. Press, 1973.

Ferguson, C. A., & Farwell, C. B. Words and sounds in early language acquisition. *Language*, 1975, *51*, 419–439.

Fitch, H. L., Halwes, T., Erickson, D., & Liberman, A. M. Perceptual equivalence of two acoustic cues for stop-consonant manner. *Perception and Psychophysics*, 1980, *27*, 343–350.

Flanagan, J. L. *Speech analysis, synthesis and perception*. (2nd ed.). New York: Academic Press, 1973.

Fodor, J. A., Garrett, M. F., & Shapero, D. B. Discriminations among phones by infants. *Quarterly Progress Report: Research Laboratory of Electronics (M.I.T.)*, 1970, *96*, 180.

Fodor, J. A., Garrett, F. M., & Brill, S. L. Pi ka pu: The perception of speech sounds by prelinguistic infants. *Perception and Psychophysics*, 1975, *18*, 74–78.

Fowler, C. A. Coarticulation and theories of extrinsic timing. *Journal of Phonetics*, 1980, *8*, 113–133.

Fowler, C. A., & Shankweiler, D. P. Identification of vowels in speech and nonspeech contexts. *Journal of the Acoustical Society of America*, 1978, *63*, S4. (Supplement)

Fowler, C. A., & Turvey, M. T. Skill acquisition: An event approach with special reference to searching for the optimum of a function of several variables. In G. E. Stelmach (Ed.), *Information processing in motor control and learning*. New York: Academic Press, 1978.

Fowler, C. A., Rubin, P., Remez, R. E., & Turvey, M. T. Implications for speech production of a general theory of action. In B. Butterworth (Ed.), *Speech production*. New York: Academic Press, 1980.

Galunov, V. I., & Chistovich, L. A. Relationship of motor theory to the general problem of speech recognition. *Soviet Physics-Acoustics*, 1966, *11*, 357–365. (Review)

Garnica, O. The development of phonemic speech perception. In T. E. Moore (Ed.), *Cognitive development and the acquisition of language*. New York: Academic Press, 1973.

Gibson, E. J. How perception really develops: A view from outside the network. In D. LaBerge & S. J. Samuels (Eds.), *Basic processes in reading: Perception and comprehension*. Potomac, Md.: Erlbaum, 1977.

Gibson, E. J. *Infants' perception of structural invariants of objects*. Psychology Department, Columbia University, New York, Spring 1980.

Gibson, E. J., & Levin, H. *The psychology of reading*. Cambridge: M.I.T. Press, 1975.

Gibson, J. J. *The senses considered as perceptual systems*. Boston: Houghton Mifflin, 1966.

Gibson, J. J., & Gibson, E. J. Perceptual learning: Differentiation or enrichment? *Psychological Review*, 1955, *62*, 32–41.

Glanville, B. B., Best, C. T., & Levenson, R. A cardiac measure of cerebral asymmetries in infant auditory perception. *Developmental Psychology*, 1977, *13*, 54–59.

Goldstein, U. G. Modeling children's vocal tracts. *Journal of the Acoustical Society of America*, 1979, *65*(Suppl. 1), S25. (Abstract)

Haith, M. M. Visual cognition in early infancy. In R. B. Kearsley & I. E. Sigel (Eds.), *Infants at risk: Assessment of cognitive function.* Hillsdale, N.J.: Erlbaum, 1979.

Hillenbrand, J., Minifie, F. D., & Edwards, T. J. Tempo of spectrum change as a cue in speech-sound discrimination by infants. *Journal of Speech and Hearing Research,* 1979, *22,* 147–165.

Holmberg, T. L., Morgan, K. A., & Kuhl, P. K. *Speech perception in early infancy: Discrimination of fricative consonants.* Paper presented at the meeting of the Acoustical Society of America, December 1977.

Horridge, G. A. *Interneurons.* London: W. H. Freeman, 1968.

Howell, P. Identification of vowels in and out of context. *Journal of the Acoustical Society of America,* 1981, *70,* 1256–1260.

Jacob, F. Evolution and tinkering. *Science,* 1977, *196,* 1161–1166.

Jacobson, M. *Developmental neurobiology.* New York: Plenum Press, 1978.

Jenkins, J. J. Remember that old theory of memory? Well, forget it! *American Psychologist,* 1974, *29,* 785–795.

Jenkins, J. J. Research in child phonology: Comments, criticism, and advice. In G. H. Yeni-Komshian, J. F. Kavanaugh, & C. A. Ferguson (Eds.), *Child phonology: Vol. 2: Perception.* New York: Academic Press, 1980.

Jusczyk, P. W. Perception of syllable-final stop consonants by two-month-old infants. *Perception and Psychophysics,* 1977, *21,* 450–454.

Jusczyk, P. W., & Thompson, E. Perception of a phonemic contrast in multisyllabic utterances by 2-month-old infants. *Perception and Psychophysics,* 1978, *23,* 105–109.

Jusczyk, P. W. Infant speech perception: A critical appraisal. In P. D. Eimas & J. L. Miller (Eds.), *Perspectives on the study of speech.* Hillsdale, N.J.: Erlbaum, 1981. (a)

Jusczyk, P. W. The processing of speech and nonspeech sounds by infants: Some implications. In R. N. Aslin, J. R. Alberts, & M. R. Petersen (eds.), *Development of perception: Psychobiological perspectives.* (Vol. 1.) New York: Academic Press, 1981 (b).

Jusczyk, P. W., Rosner, B. S., Cutting, J. E., Foard, C. F., & Smith, L. B. Categorical perception of nonspeech sounds by 2-month-old infants. *Perception and Psychophysics,* 1977, *21,* 50–54.

Jusczyk, P. W. Murray, J., & Bayly, J. *The perception of place of articulation of fricatives and stops by infants.* Paper presented at the meeting of the Society for Research on Child Development, San Francisco, March 1979.

Jusczyk, P. W., Pisoni, D. B., Walley, A., & Murray, J. Discrimination of relative onset time of two-component tones by infants. *Journal of the Acoustical Society of America,* 1980, *67,* 262–270.

Kaye, K. Why we don't talk "baby talk" to babies. *Journal of Child Language,* 1980, *7,* 489–507.

Kent, R. D., & Forner, L. L. Developmental study of vowel formant frequencies in an imitation task. *Journal of the Acoustical Society of America,* 1979, *65,* 208–217.

Kimura, D. The asymmetry of the human brain. *Scientific American,* 1973, *228,* 70–78.

Kinsbourne, M. The control of attention by interaction between the hemispheres. In S. Kornblum (Ed.), *Attention and performance IV.* New York: Academic Press, 1973.

Kinsbourne, M. Hemispheric specialization and the growth of human understanding. *American Psychologist,* 1982, *37,* 411–420.

Kuhl, P. K. Predispositions for perception of speech-sound categories: A species-specific phenomenon? In F. D. Minifie & L. L. Lloyd (Eds.), *Communicative and cognitive abilities—Early behavioral assassment.* Baltimore, Md.: University Park Press, 1978.

Kuhl, P. K. Speech perception in early infancy: Perceptual constancy for spectrally dissimilar vowel categories. *Journal of the Acoustical Society of America,* 1979, *66,* 1668–1679.

Kuhl, P. K. Perceptual constancy for speech-sound categories in early infancy. In G. H. Yeni-Komshian, J. F. Kavanaugh & C. A. Ferguson (Eds.), *Child phonology: Vol. 2: Perception.* New York: Academic Press, 1980.

Kuhl, P. K. Auditory category formation and developmental speech perception. In R. E. Stark (Ed.), *Language behavior in infancy and early childhood.* New York: Elsevier-North Holland, 1981. (a)

Kuhl, P. K. Discrimination of speech by nonhuman animals: Basic auditory sensitivities conducive to the perception of speech-sound categories. *Journal of the Acoustical Society of America*, 1981. (b)

Kuhl, P. K., & Meltzoff, A. N. Bimodal speech perception in early infancy. *Journal of the Acoustical Society of America*, 1982, *71*, (Suppl.) S77–78. (Abstract)

Kuhl, P. K., & Miller, J. D. Speech perception by the chinchilla: Voiced-voiceless distinction in alveolar plosive consonants. *Science*, 1975, *190*, 69–72.

Kuhl, P. K., & Miller, J. D. Speech perception by the chinchilla: Identification functions for synthetic VOT stimuli. *Journal of the Acoustical Society of America*, 1978, *63*, 905–917.

Lehiste, I., & Meltzer, D. Vowel and speaker identification in natural and synthetic speech. *Language and Speech*, 1973, *16*, 356–364.

Lenneberg, E. H. *Biological foundations of language.* New York: Wiley, 1967.

Liberman, A. M. On finding that speech is special. *American Psychologist*, 1982, *37*, 148–167.

Liberman, A. M., Cooper, F. S., Shankweiler, D., & Studdert-Kennedy, M. The speech code. *Psychological Review*, 1967, *74*, 431–461.

Lieberman, P. *Intonation, perception and language.* Cambridge: M.I.T. Press, 1967.

Lieberman, P. On the development of vowel production in young children. In G. H. Yeni-Komshian, J. F. Kavanaugh, & C. A. Ferguson (Eds.), *Child phonology. Vol. 1: Production.* New York: Academic Press, 1980.

Lieberman, P., Harris, K. S., Wolff, P., & Russell, L. H. Human infant cry and nonhuman primate vocalization. *Journal of Speech and Hearing Research*, 1971, *14*, 718–727.

Lisker, L. Rapid vs. rabid: A catalogue of acoustic features that may cue the distinction. *Haskins Laboratories: Status Report on Speech Research*, 1978, *SR-54*, 127–132.

MacDonald, J., & McGurk, H. Visual influences on speech perception processes. *Perception and Psychophysics*, 1978, *24*, 253–257.

MacKain, K. S., Studdert-Kennedy, M., Spieker, S., & Stern, D. *Cross-modal coordination in infants' perception of speech.* Paper presented at the meeting of the 2nd International Conference on Child Psychology, Vancouver, August 1981

MacKain, K. S., Studdert-Kennedy, M., Spieker, S., & Stern, D. Infant intermodal speech perception is a left hemisphere function. *Science*, 1983, *219*, 1347–1349.

Malsheen, B. J. Two hypotheses for phonetic clarification in the speech of mothers to children. In G. H. Yeni-Komshian, J. F. Kavanaugh, & C. A. Ferguson (Eds.), *Child phonology. Vol. 2: Perception.* New York: Academic Press, 1980.

Mann, V. A. Influence of preceding liquid on stop consonant perception. *Perception and Psychophysics*, 1980, *28*, 407–412.

Mann, V. A., & Repp, B. H. Influence of vocalic context on perception of the esh-/s/ distinction. *Perception and Psychophysics*, 1980, *28*, 213–228.

Mann, V. A., Madden, J., Russell, J. M., & Liberman, A. M. Further investigation into the influence of preceding liquids on stop consonant perception. *Journal of the Acoustical Society of America*, 1981, *69*(Suppl.), S91 (A).

Marshall, W. A. *The development of the brain.* London: Oliver & Boyd, 1968.

McCulloch, W. S. *The embodiments of mind.* Cambridge: M.I.T. Press, 1965.

Mehler, J., & Bertoncini, J. Infants' perception of speech and other sounds. In J. Morton & J. Marshall (Eds.), *Psycholinguistics. Series II.* London: Elek Science, 1978.

Menn, L. Phonological theory and child phonology. In G. H. Yeni-Komshian, J. F. Kavanaugh, & C. A. Ferguson (Eds.), *Child phonology. Vol. 1: Production.* New York: Academic Press, 1980.

Menyuk, P., & Menn, L. Early strategies for the perception and production of words and sounds. In P. Fletcher & M. Garman (Eds.), *Studies in language acquisition*. Cambridge, England: Cambridge University Press, 1979.

Miller, C. L., & Morse, P. A. The "heart" of categorical speech discrimination in young infants. *Journal of Speech and Hearing Research*, 1976, *19*, 578–589.

Miller, C. L., Morse, P. A., & Dorman, M. F. Cardiac indices of infant speech perception: Orienting and burst discrimination. *Quarterly Journal of Experimental Psychology*, 1977, *29*, 533–545.

Miller, C. L., Younger, B. A., & Morse, P. A. The categorization of male and female voices in infancy. *Infant Behavior and Development*, 1982, *5*, 143–159.

Miller, J. D. Perception of speech sounds in animals: Evidence for speech processing by mammalian auditory mechanisms. In T. H. Bullock (Ed.), *Recognition of complex acoustic signals*. Berlin: Abakon Verlagsgesellschaft, 1977.

Miller, J. D., Wier, C. C., Pastore, R., Kelly, W. J., & Dooling. R. J. Discrimination and labeling of noise-buzz sequences with varying noise-lead times: An example of categorical perception. *Journal of the Acoustical Society of America*, 1976, *60*, 410–417.

Miller, J. D., Henderson, B. C., Sullivan, H. T., & Rigden, G. K. Speech perception by the chinchilla: Learning functions for pairs of synthetic stimuli from various points along a VOT continuum. *Journal of the Acoustical Society of America*, 1978, *64* (Suppl.), S18 (A).

Miller, J. L., & Eimas, P. D. *Contextual perception of voicing by infants*. Paper presented at the meeting of the Society for Research in Child Development, Boston, April 1981.

Moffitt, A. R. Consonant cue perception by twenty- to twenty-four week old infants. *Child Development*, 1971, *42*, 717–731.

Molfese, D. L., & Molfese, V. J. VOT distinctions in infants: Learned or innate? In H. A. Whitaker & H. Whitaker (Eds.), *Studies in neurolinguistics*. (Vol. 4.) New York: Academic Press, 1979.

Molfese, D. L., Freeman, R. B., & Palermo, D. S. The ontogeny of brain lateralization for speech and nonspeech stimuli. *Brain and Language*, 1975, *2*, 356–368.

Molfese, D. L., Nuñez, V., Seibert, S. M., & Ramaniah, N. V. Cerebral asymmetry: Changes in factors affecting its development. *Annals of the New York Academy of Science*, 1976, *280*, 821–833.

Morais, J., Cary, L., Alegria, J., & Bertelson, P. Does awareness of speech as a sequence of phones arise spontaneously? *Cognition*, 1979, *7*, 323–331.

Morse, P. A. The discrimination of speech and nonspeech stimuli in early infancy. *Journal of Experimental Child Psychology*, 1972, *14*, 477–492.

Morse, P. A. Infant speech perception: Origins, processes, and Alpha Centauri. In F. Minifie, & L. L. Lloyd (Eds.), *Communicative and cognitive abilities—Early behavioral assessment*. Baltimore, Md.: University Park Press, 1978.

Morse, P. A., Eilers, R. E., & Gavin, W. J. The perception of the sounds of silence in early infancy. *Child Development*, 1982, *53*, 189–195.

Morse, P. A., & Snowden, C. T. An investigation of categorical speech discrimination by rhesus monkeys. *Perception and Psychphysics*, 1975, *17*, 9–16.

Neuweiler, G., Bruns, V., & Schuller, G. Ears adapted for the detection of motion, or how echolocating bats have exploited the capacities of the mammalian auditory system. *Journal of the Acoustical Society of America*, 1980, *68*, 741–753.

Oller, D. K. The emergence of the sounds of speech in infancy. In G. H. Yeni-Komshian, J. F. Kavanaugh, & C. A. Ferguson (Eds.), *Child phonology. Vol. 1: Production*. New York: Academic Press, 1980.

Paget, R. *Human speech*. London: Routledge & Keegan-Paul, 1930.

Petersen, M. R. Perception of acoustic communication signals by animals: Developmental perspectives and implications. In R. N. Aslin, J. R. Alberts, & M. R. Petersen (Eds.), *Development of perception: Psychobiological perspectives.* (Vol. 1.) New York: Academic Press, 1981.

Pike, K. L. Language as particle, wave, and field. *The Texas Quarterly,* 1959, *2,* 37–55.

Pisoni, D. B. Auditory and phonetic memory codes in the discrimination of consonants and vowels. *Perception and Psychophysics,* 1973, *13,* 253–260.

Pisoni, D. B. Auditory short term memory and vowel perception. *Memory and Cognition,* 1975, *3,* 7–18.

Pisoni, D. B. Identification and discrimination of the relative onset of two-component tones: Implications for the perception of voicing in stops. *Journal of the Acoustical Society of America,* 1977, *61,* 1352–1361.

Pisoni. D. B. Speech perception. In W. K. Estes (Ed.), *Handbook of cognitive processes. Vol. 6: Linguistic functions in cognitive theory.* Hillsdale, N.J.: Erlbaum, 1978.

Rabinowicz, T. The differentiate maturation of the human cerebral cortex. In F. Falkner & J. M. Tanner (Eds.), *Human growth: Vol. 3: Neurobiology and nutrition.* New York: Plenum Press, 1979.

Rakerd, B., & Verbrugge, R. R. Individual differences in the perception of isolated vowels and vowels in a consonantal context. *Journal of the Acoustical Society of America,* 1982, *71,* S76 (A).

Repp, B. H. Categorical perception. *Haskins Laboratories: Status Report on Speech Research,* 1982, *SR-70,* 99–183.

Riedel, K. *Durational factors in the speech perception of aphasics.* Unpublished doctoral dissertation, City University of New York Graduate Center, 1981.

Robson, R. C., Morrongiello, B. A., Best, C. T., & Clifton, R. K. Trading relations in the perception of speech by five-year-old children. *Haskins Laboratories: Status Report on Speech Research,* 1982, *SR-71/72,* in press.

Rose, S. R. *The conscious brain.* New York: Vintage Books, 1976.

Rosen, S. M., & Howell, P. Plucks and bows are not categorically perceived. *Perception and Psychophysics,* 1981, *30,* 156–168.

Ruff, H. A. The development of perception and recognition of objects. *Child Development,* 1980, *51* 981–992.

Ruff, H. A. Effect of object movement on infants' detection of object structure. *Developmental Psychology,* 1982, *18,* 462–472.

Schubert, E. D. The role of auditory perception in language processing. In D. D. Duané & M. B. Rawson (Eds.), *Reading, perception and language.* Baltimore: York Press, 1974.

Schvachkin, N. K. The development of phonemic speech perception in early childhood. In C. A. Ferguson & D. I. Slobin (Eds.), *Studies of child language development.* New York: Holt, Rhinehart & Winston, 1973.

Schwartz, J., & Tallal, P. Rate of acoustic change may underlie hemispheric specialization for speech perception. *Science,* 1980, *207,* 1380–1381.

Searle, C. L., Jacobson, J. Z., & Rayment, S. G. Phoneme recognition based on human audition. *Journal of the Acoustical Society of America,* 1979, *65,* 799–809.

Shankweiler, D. P., Strange, W., & Verbrugge, R. R. Speech and the problem of perceptual constancy. In R. E. Shaw & J. Bransford (Eds.), *Perceiving, acting, and knowing: Toward an ecological psychology.* Hillsdale, N.J.: Erlbaum, 1977.

Shaw, R. E., & Pittenger, J. Perceiving the face of change in changing faces: Implications for a theory of object perception. In R. E. Shaw & J. Bransford (Eds.), *Perceiving, acting, and knowing: Toward an ecological psychology.* Hillsdale, N.J.: Erlbaum, 1977.

Simon, C., & Fourcin, A. J. Cross-language study of speech-pattern learning. *Journal of the Acoustical Society of America,* 1978, *63,* 925–935.

Sinnott, J. M., Beecher, M. D., Moody, D. B., & Stebbins, W. C. Speech sound discrimination by monkeys and humans. *Journal of the Acoustical Society of America*, 1976, *60*, 687–695.

Spelke, E. S. Exploring audible and visible events in infancy. In A. D. Pick (Ed.), *Perception and its development: A tribute to Eleanor J. Gibson*. Hillsdale, N.J.: Erlbaum, 1979.

Spelke, E. S., & Cortelyou, A. Perceptual aspects of social knowing. In M. Lamb & E. Sherrod (Eds.), *Infant social cognition*. Hillsdale, N.J.: Erlbaum, 1981.

Stark, R. E. Stages of speech development. In G. H. Yeni-Komshian, J. F. Kavanaugh, & C. A. Ferguson (Eds.), *Child phonology. Vol. 1: Production*. New York: Academic Press, 1980.

Stern, D. N., Spieker, S., & MacKain, K. S. Intonation contours as signals in maternal speech to infants. *Developmental Psychology*, 1982, *18*, 717–735.

Stevens, K. N. The quantal nature of speech perception: Evidence from articulatory-acoustic data. In E. E. David & P. P. Denes (Eds.), *Human communication: A unified view*. New York: McGraw Hill, 1972.

Stevens, K. N., & Blumstein, S. E. Invariant cues for place of articulation in stop consonants. *Journal of the Acoustical Society of America*, 1978, *64*, 1358–1368.

Strange, W., Verbrugge, R. E., Shankweiler, D. P., & Edman, T. R. Consonant environment specifies vowel identity. *Journal of the Acoustical Society of America*, 1976, *60*, 213–224.

Strange, W., Jenkins, J. J., & Edman, T. R. *Journal of the Acoustical Society of America*, 1977, *61*, S39 (A).

Streeter, L. A. Language perception of 2-month-old infants shows effects of both innate mechanisms and experience. *Nature*, 1976, *259*, 39.

Studdert-Kennedy, M. The emergence of phonetic structure. *Cognition*, 1981, *10*, 301–306. (a)

Studdert-Kennedy, M. A note on the biology of speech perception. *Haskins Laboratories: Status Report on Speech Research*, 1981, *SSR-65*, 223–232. (b)

Studdert-Kennedy, M. Cerebral hemispheres: Specialized for the analysis of what? *Behavioral and Brain Sciences*, 1981, *4*, 76–77. (c)

Studdert-Kennedy, M., Perceiving phonetic segments. In T. Myers, J. Laver, & J. Anderson (Eds.), *The cognitive representation of speech*. Amsterdam: North-Holland, 1981. (d)

Studdert-Kennedy, M., & Lane, H. The structuring of language: Clues from the differences between signed and spoken language. *Haskins Laboratories: Status Report on Speech Research*, 1980, *SR-63/64*, 1–8.

Studdert-Kennedy, M. & Shankweiler, D. P. Hemispheric specialization for speech perception. *Journal of the Acoustical Society of America*, 1970, *48*, 579–594.

Studdert-Kennedy, M., & Shankweiler, D. P. Hemispheric specialization for language processes. *Science*, 1980, *211*, 960–961.

Studdert-Kennedy, M., Liberman, A. M., Harris, K. S., & Cooper, F. S. Motor theory of speech perception: A reply to Lane's critical review. *Psychological Review*, 1970, *77*, 234–249.

Summerfield, Q. Perceptual learning and phonetic perception. In *Interrelations of the communicative senses: Proceedings of the National Science Foundation conference at Asilomar, California, September–October 1978*. Washington, D. C.: National Science Foundation Publications, 1978.

Summerfield, Q. Use of visual information for phonetic perception. *Phonetica*, 1979, *36*, 314–331.

Summerfield, Q. Differences between spectral dependencies in auditory and phonetic temporal processing: Relevance to the perception of voicing in initial stops. *Journal of the Acoustical Society of America*, 1982, *72*, 51–61.

Summerfield, Q., McGrath, M., & Forster, J. *Detection and resolution of audio-visual conflict in the perception of vowels*. Paper presented at the meeting of the Acoustical Society of America, Chicago, April 1982.

Swoboda, P. J., Kaas, J., Morse, P. A., & Leavitt, L. A. Memory factors in infant vowel discrimination in normal and at-risk infants. *Child Development*, 1978, *49*, 332–339.

Swoboda, P. J., Morse, P. A., & Leavitt, L. A. Continuous vowel discrimination in normal and at-risk infants. *Child Development*, 1976, *47*, 459–465.

Trehub, S. E. Infants' sensitivity to vowel and tonal contrasts. *Developmental Psychology*, 1973, *9*, 91–96.

Trehub, S. E. & Rabinovitch, M. S. Auditory-linguistic sensitivity in early childhood. *Developmental Psychology*, 1972, *6*, 74–77.

Trehub, S. E., Bull, D., & Schneider, B. A. Infant speech and nonspeech perception: A review and reevaluation. In R. L. Schiefelbusch & D. D. Bricker (Eds.), *Early language acquisition and intervention: Language intervention series*. Baltimore, Md.: University Park Press, 1981.

Trevarthen, C. Neuroembryology and the development of perception. In F. Falkner & J. M. Tanner (Eds.), *Human growth: Neurobiology and nutrition*. New York: Plenum Press, 1979.

Tuchmann-Duplessis, H., Aroux, M., & Haegel, P. *Illustrated human embryology. Vol. 3: Nervous system and endocrine glands*. New York: Springer-Verlag, 1975.

Verbrugge, R. R., & Rakerd, B. Talker-independent information for vowel identity. *Haskins Laboratories: Status Report on Speech Research*, 1980, *SR-62*, 205–215.

Verbrugge, R. R., Shankweiler, D. P., & Fowler, C. A. Context-conditioned specification of vowel identity. *Haskins Laboratories: Status Report on Speech Research*, 1980, *SR-62*, 217–224.

Verbrugge, R. R., Rakerd, B., Fitch, H., Tuller, B., & Fowler, C. A. The perception of speech events: An ecological perspective. In R. E. Shaw & W. Mace (Eds.), *Event perception*. Hillsdale, N.J.: Erlbaum, in press.

Walley, A. C., Pisoni, D. B., & Aslin, R. N. The role of early experience in the development of speech perception. In R. N. Aslin, J. R. Alberts, & M. R. Petersen (Eds.), *Development of perception: Psychobiological perspectives. Vol. 1*. New York: Academic Press, 1981.

Warren, W. H., & Verbrugge, R. R. Toward an ecological acoustics. In R. E. Shaw & W. Mace (Eds.), *Event perception*. Hillsdale, N.J.: Erlbaum, in press.

Waters, R. S., & Wilson, W. A. Speech perception by rhesus monkeys: The voicing distinction in synthesized labial and velar stop consonants. *Perception and Psychophysics*, 1976, *19*, 285–289.

Weiss, M. S., & House, A. S. Perception of dichotically presented vowels. *Journal of the Acoustical Society of America*, 1973, *53*, 51–53.

Werker, J. F. *Language-specific developmental changes in the speech perception of prelinguistic infants*. Paper presented at the meeting of the Society for Research in Child Development, Detroit, April 1983.

Werker, J. F., & Tees, R. C. *Developmental changes in the perception of non-native speech sounds*. Papers presented at the meeting of the Canadian Psychological Association, Toronto, June 1981.

Werker, J. F., & Tees, R. C. *Evidence for perceptual reorganization during the first year of life*. Paper presented at the meeting of the Canadian Psychological Association, Montreal, June 1982.

Williams, L., & Bush, M. Discrimination by young infants of voiced stop consonants with and without release bursts. *Journal of the Acoustical Society of America*, 1978, *63*, 1223–1226.

Wood, C. C. Discriminability, response bias, and phoneme categories in discrimination of voice onset time. *Journal of the Acoustical Society of America*, 1976, *60*, 1381–1389.

Zlatin, M. & Koenigsknecht, R. Development of the voicing contrast: A comparison of voice onset time in stop perception and production. *Journal of Speech and Hearing Research*, 1976, *19*, 78–92.

Zoloth, S. R., Petersen, M. R., Beecher, M. D., Green, S., Marler, P., Moody, D. B., & Stebbins, W. Species-specific perceptual processing of vocal sounds by monkeys. *Science*, 1979, *204*, 870–873.

The Socio-emotional Development of Maltreated Children
An Empirical and Theoretical Analysis

J. LAWRENCE ABER AND DANTE CICCHETTI

1. INTRODUCTION

The thought of a maltreated child conjurs up images of bruises, fractures, malnutrition, and the like. But it is the emotional damage, not the physical damage, that, despite the difficulty in operationalizing the concept and despite the legal and political controversy surrounding the concept, may have the most frequent, long-term deleterious effect on the development of maltreated children (Juvenile Justice Standards Project, 1977). It is curious, then, that among the many issues concerning the phenomenon of child maltreatment, the issue of the impact of maltreatment on a child's socio-emotional development took so long to interest researchers.

Empirical research on the socio-emotional development sequelae of child maltreatment is a recent phenomenon. Only in the past decade has the impact of maltreatment on children's socio-emotional development become the focus of systematic psychological study. Studies of the sequelae of maltreatment are very important for enhancing the quality of clinical, legal, and policy-making decisions for maltreated children. Decisions concerning such

J. Lawrence Aber • Department of Psychology, Barnard College, Columbia University, New York, New York 10027. Dante Cicchetti • Department of Psychology and Social Relations, Williams James Hall, Harvard University, Cambridge, Massachusetts 02138.

issues as whether to report a child as maltreated, whether to remove coercively a child from the home, how to develop services to meet the specific psychological needs of maltreated children, and how to evaluate these services would all benefit from a more solid and sophisticated data base on the socioemotional sequelae of maltreatment (Aber, 1980; Aber & Zigler, 1981).

The basic elements of empirical studies of the socio-emotional sequelae of child maltreatment differ greatly from study to study. The research *subjects* of studies of child maltreatment vary in age, subtypes of maltreatment suffered, and many important demographic characteristics such as socio-economic status. The *theories* guiding the precise questions posed by the studies also vary greatly. Some studies are virtually nontheoretical whereas others are implicitly or explicitly wed to a major theoretical orientation like psychoanalysis or behaviorism. Some studies focus on stage-related developmental issues, others on nonstage-related enduring psychological traits. Some studies are guided by theories of normal development and some are not. Perhaps most importantly, there has been great variety in the experimental *design* and research *methods* employed by various scholars, which in turn greatly influence the inferences one can draw from the findings of these studies. Although the logic and validity of inferences drawn from psychological studies are always a matter of scientific concern, they also become a matter of great practical concern when the inferences may influence clinical, legal, and policy decisions about maltreated children.

This chapter will critically review the last two decades of empirical research on the socio-emotional development of maltreated children. The approach to the review will be an in-depth analysis of a large and representative sample of the most influential studies conducted during this time. The review will focus on how variations in subjects, underlying theories, and experimental designs, and research methods affect the conclusions one can draw from the last two decades of research. Throughout this chapter, we critique the methods employed by various investigators of the socio-emotional development of maltreated children. We have resisted the temptation to write a single section on general methodological limitations of this corpus of studies and instead have critiqued the research methods of each study individually. We have adopted this strategy in order to provide methodologically inexperienced readers with a more detailed picture of precisely how the findings of specific studies are challenged by their methodological limitations. We know of no better heuristic device to convey this critical message than to confront the limits of each research study.

Given the great importance of design and methodological issues in influencing the conclusions drawn from the research, the chapter will be organized into five major sections, each devoted to a different type of design for the empirical study of the socio-emotional developmental sequelae of maltreatment.

Clinical studies dominated the research on the socio-emotional sequelae of maltreatment from the mid-1960s to the early 1970s and are still common in the literature today. Clinical studies were and are valuable in generating hypotheses about the possible effects of maltreatment on child development. But the diversity of clinical descriptions of maltreated children, many seemingly at odds with each other (e.g., are maltreated children excessively wary of or excessively friendly to strangers?), led to a perceived need for more empirical studies of the socio-emotional developmental sequelae of maltreatment. Consequently, *cross-sectional empirical studies* began to be mounted in the late 1960s. At first these studies did not always include comparison groups and the validity and generalizability of the findings were questioned. Later cross-sectional studies have paid more careful attention to the issues of appropriate comparison groups. In the early 1970s, a growing awareness of the importance of the long-term nature of the impairments in development due to child maltreatment stimulated a series of *follow-up studies*. Follow-up studies attempted to distinguish between transient and more enduring effects of child maltreatment.

Despite the methodological improvements made by cross-sectional and follow-up studies over strictly clinical descriptions, the cumulative results of the studies were still difficult to integrate into a coherent picture of the socio-emotional sequelae of child maltreatment. This was due in large measure to the absence of an overarching, integrative developmental theory to guide the design of studies and the interpretation of findings. Recently, however, a number of *theoretically-derived studies* of the socio-emotional sequelae of child maltreatment have begun to appear. Theoretically, these studies have adopted a truly developmental perspective by specifying stage-salient developmental competencies that may be at risk due to a history of maltreatment. They have also made important methodological contributions through the use of age-appropriate experimental tasks and assesment procedures to assess those competencies (Sroufe, 1979a,b). The adoption of an overarching, integrative developmental theory has contributed to an increased awareness of the need for *longitudinal prospective studies* of the development of maltreated children in order to disentangle cause from effect and to specify developmental processes influenced by maltreatment.

After in-depth analysis of these five types of studies (clinical, cross-sectional, follow-up, theoretically-derived, and longitudinal), the concluding section of this chapter will make a series of recommendations for improving the quality of future research on the socio-emotional development of maltreated children. These recommendations are not exhaustive, but instead represent our judgment of those improvements that are most important scientifically and practically.

2. STUDIES OF THE SOCIO-EMOTIONAL DEVELOPMENT OF MALTREATED CHILDREN

2.1. Clinical Observations and Descriptions of the Socio-emotional Development of Maltreated Children

Clinical observations suggest that maltreated children exhibit a variety of clinical symptoms presumed to indicate aberrant emotional development. These studies chiefly consist of characterizations of maltreated children seen in child protection units or in conjunction with treatment programs for families receiving child protection. In the vast majority, no control groups have been employed. These clinical descriptions have been gathered in diverse settings (e.g., upon admission into hospital emergency rooms, during home visits with mothers, in treatment centers, in special protective day-care centers, during therapy sessions), by a variety of professional personnel (e.g., social workers, emergency room physicians and nurses, pediatricians, psychologists, teachers, and psychiatrists), utilizing different means of data collection (e.g., maternal report during home interviews, observing play therapy sessions, asking teachers to make behavioral ratings, social worker impressions based on case record data, home interviews and observations, nursing staff descriptions of behavior on hospital wards, and watching children interact with adults and peers in day-care and school settings). Based on these sources of data, it appears that the social/emotional consequences of child maltreatment are variable.

Regardless of the source of the information, maltreated children have been portrayed as whiny, fussy, chronically crying, unresponsive, passive, compliant, listless, apathetic, fearful, panicky, negativistic, uncontrollable, unsmiling, and unappealing (Galdston, 1965; Johnson & Morse, 1968). Other investigators have described maltreated children as manifesting inappropriate positive affect (such as smiling at the beginning of a medical procedure), being indiscriminately friendly, and/or being totally unresponsive to others, demonstrating a lack of overt emotional expressiveness (Galdston, 1975; Gray & Kempe, 1976; Martin & Beezeley, 1977; Terr, 1970), as demonstrating compliance, hypervigilance, "frozen watchfulness," inhibition, minimal expression of negative affect or pain, extreme withdrawal, depression, anhedonia, lack of impulse control; and exhibiting emotional disturbance and behavior problems at home and at school (Elmer, 1967; Martin &

Beezeley, 1977; Morse, Sahler, & Friedman, 1970; Ounsted, Oppenheimer, & Lindsay, 1974; Rolston, 1971). Additional investigators have described maltreated children as violent and aggressive, extremely resistant, angry, and full of hatred (Fontana, 1973; Galdston, 1971, 1975; Green, 1978a; Kempe & Kempe, 1978), uncuddly (Ounsted et al., 1974), anxious, hostile, and low in empathy (N. D. Feshbach, 1973), with an associated inability to trust others, develop good object relations, form a positive sense of self, interact harmoniously with others, and form a self-identity (Cohn, 1979; Green, 1978a; Kempe & Kempe, 1978; Ounsted et al., 1974).

Two clinical reports will now be discussed in detail to offer the reader a sense of the nature of these investigations. Terr (1970) presented clinical impressions from 10 "suspected" child-abuse cases that she evaluated over a 6-year period in her psychiatric practice. Her conclusions were based on interview and observational data, although the observations of mother-child interaction were not collected in a systematic fashion (e.g., independent, "blind" raters were not used, only six mother-"abused" child dyads were observed, and in three cases mothers were observed with another sibling). The ten youngsters ranged in age from three months to nine years. Each had experienced multiple injuries prior to the "suspected" abuse report. The families were very heterogeneous socio-economically. Three had not graduated from high school, three were high school graduates, three attended college for some period of time, and one had a doctorate degree.

Terr (1970) reported a variety of ego defects in these children, with symptoms of ego disturbance varying from each age group. For example, the abused infants exhibited withdrawal, indifference to the mother, and psychomotor retardation. Ego defects of older children consisted of shallow relationships with the parents and a lack of differentiation among relationships to others (e.g., superficial attachment of an indiscriminatory sort). Terr hypothesized that the presence of ego defects in "abused" children is related to the extreme dependence on the parent that the child has during the infancy and early childhood periods. However, these results must be interpreted with caution for a variety of reasons. First, no control group of nonabusing families was involved. Second, the cases were not legally verified child-maltreatment families. Finally, the same rater, cognizant of the diagnostic status of the families, conducted the interviews and observations.

Based on the observations of 20 physically abused children seen in outpatient psychotherapy twice weekly for a year, Green (1978a) described these maltreated children as sad and dejected, self-deprecatory, and low in self-esteem and frustration tolerance. Furthermore, Green (1978a) noted that these children often acted in a provocative, belligerent, "limit-testing" fashion: exactly the type of behavior that could elicit severe parental punishment.

This study illustrates the value of a clinical approach for identifying potentially important aspects of maltreated children's socio-emotional development. However, the lack of an appropriate control group raises questions about whether these characteristics are true of maltreated children specifically or of all clinically referred children in general. Moreover, the restriction of observations to psychotherapy sessions, makes it impossible to know whether these characteristics would occur in the enduring natural environs of the child's life (e.g., at home, in school, or with the peer group) or are limited to the more rarefied atmosphere of an intensive psychotherapeutic relationship.

In another paper, Green (1978b) reported that abused children often engaged in overt types of self-destructive behavior, such as self-mutilation and suicide attempts, gestures, and threats. Green (1978b) studied 60 documented physically abused children who were also neglected and compared them to 30 confirmed nonabused neglect cases and 30 control cases. These children ranged in age between 5 and 13 years, with an average age of 8½ years. All children came from lower income families and the majority lived at home. Based on an interview with either the mother or maternal guardian of these children, Green found a significantly higher incidence of self-destructive behavior in the physically abused children. Of the abused children, 41% engaged in self-destructive acts, the majority occurring either after parental physical punishment or in response to actual or threatened separation or abandonment from their parents or caretakers. Concurrent psychological assessment of these children ruled out brain damage or psychosis as an explanation of the results. No significant differences were found between intelligence test profiles of self-destructive abused children and their non-self-destructive abused counterparts. These clinical findings extend Green's earlier report of a link between physical abuse and subsequent self-destructive behavior in schizophrenic children (Green, 1968).

2.2. Cross-sectional Studies of the Socio-emotional Consequences of Maltreatment

Harold Martin and his colleagues at the University of Colorado Medical center have been among the foremost proponents of the need to document the psychological consequences of maltreatment *for the child* (Martin, 1976; Martin & Beezley, 1976, 1977; Martin, Beezley, Conway, & Kempe, 1974). Martin *et al.* (1974) state their position quite strongly:

> The abusive environment . . . impairs the development of the child neurologically, cognitively, and emotionally. . . . The child should not return to a home in which neglect, undernutrition and deprivation will continue to thwart his future development. (p. 43)

Furthermore, they assert that: "Therapy for parents, legal maneuvers, welfare agency policies and medical procedures must have as their overriding purpose the guarding of the health and development of the growing child" (p. 44).

Although it is hard to conceive that there are not negative consequences accompanying and/or associated with child maltreatment, it is important that such devastating impairments be documented in a scientifically rigorous fashion, especially because of the compelling legal, clinical, and policy decisions that hinge precipitously on such outcome data (Aber, 1980; Aber & Zigler, 1981; Bourne & Newberger, 1977; Goldstein, Freud, & Solnit, 1973, 1979; Juvenile Justice Standards Project, 1977; Wald, 1975, 1976).

How, then, did Martin and his colleagues go about reaching such powerful and controversial conclusions? Unfortunately, even though their behavioral observations and clinical insights provide us with an invaluable source of hypotheses to test, they illustrate the importance, indeed the necessity, of conducting empirically inscrutable research.

Here we will focus on the reports of the emotional characteristics of "abused" children (Martin & Beezley, 1976, 1977). In these studies, 50 physically abused children, ranging in age from 22 months to 13 years, were assessed approximately 4½ years after the abuse was initially identified. The mean age of the 50 children was 6 years, 5 months. Compared to earlier published reports on the psychological consequences of child abuse (see, for example, Elmer, 1967; Elmer & Gregg, 1967; Gregg & Elmer, 1969), these 50 children were less severely physically maltreated. There was no control group utilized in the Martin and Beezley studies.

Three clinicians comprised the research team. The observations were conducted in a research laboratory in the University Medical Center. Two members of the team observed each child's behavior behind a one-way mirror, while the other researcher conducted the interview and the intellectual, physical, and neurodevelopmental assessments. In an effort to make certain that the behavioral characteristics observed in the children were typical and not merely context dependent, parents, teachers, and social workers were interviewed subsequently about how the children behaved in other settings. Based on prior research studies and clinical experience, the authors "arbitrarily decided to concentrate on nine characteristics in these children" (Martin & Beezley, 1977, p. 19):

1. *Impaired ability for enjoyment,* displayed by 33 of the 50 children
2. *Behavioral symptoms* presumed to reflect emotional turmoil (e.g., enuresis, aggression toward and/or avoidance of peers, and socially inappropriate behavior), found in 31 of the children in the sample

3. *Low self-esteem,* exemplified by self-deprecation and poor self-confidence, shown by 26 of the 50 children
4. *Withdrawal,* manifested by 12 of the children and inferred on the basis of their extreme fearefulness to a variety of situations and their low tolerance of frustration
5. *Oppositional behavior,* characterized by aggressive or passive-aggressive resistance and an inability to accept limits set by adults, exhibited by 12 children
6. *Hypervigilance,* manifested by 11 of the sample and displayed through scanning their environment for cues, apparently a result of wariness of the external surround
7. *Compulsivity,* seen in 11 abused children
8. *Precocious or "pseudoadult" behavior,* found in 10 children, exemplified by inappropriate concerns for the emotional needs of adults
9. *School learning problems,* shown by 9 children, and often associated with aggressive behavior or withdrawal

These data suggest that abused children do not present a homogeneous personality type, but rather that the sequelae of maltreatment are as varied as are their causes. However, no attention is paid to any developmental changes that might occur in the personality characteristics of maltreated children. All ages are lumped together in a heterogeneous fashion, and no statistical analyses are performed that might suggest stage-specific emotional symptoms. Even if these analyses had been conducted, this study nonetheless suffers from its lack of a control group and from the fact that ratings are not made blind to the child's diagnostic status. Even though all three clinicians had to agree that a symptom was present for it to be included, and corroborative data were collected from the other sources noted, it is likely that observers would be impelled to see pathology in a group of physically abused children. Furthermore, since the vast majority of the behavioral characteristics reported subsume so many symptoms, the absence of a control group and appropriate developmental base-rates makes the data impossible to interpret. Although we feel that the preponderance of such clinical symptomatology merits significant attention regardless of whether they occur more often in abused children than in controls, if we are to ascertain the effects that maltreatment *per se* has on development, then we can no longer tolerate such egregious omissions.

In a better controlled study, Reidy (1977) examined the aggressive characteristics of 20 physically abused, 16 nonabused-neglected, and 22 non-maltreated children. The diagnostic status of the two experimental groups of maltreated children was confirmed by state programs and case workers.

The control group was referred from a day-care center serving low- and middle-income families. The three groups of children did not differ on any sociodemographic characteristics. Moreover, none of the experimental groups had any known organic damage.

At the time of the study, 12 of the abuse sample and 4 of the neglect group resided in foster-care residences. The remaining 8 abuse and 12 neglect cases lived at home. The average ages of the abuse, neglect, and control group were 6½, 7, and 6½ years, respectively. Reidy did not specify how much time had elapsed since the legal incident of maltreatment had been reported for the two experimental groups.

All children were given six cards from the Thematic Apperception Test (TAT) and asked to tell a story about each picture. These cards were employed as a putative measure of aggressive imagery. Immediately after the TAT assessment, each child was brought into a laboratory playroom at the University Mental Health Center. There they played for 20 minutes with the same examiner who administered the TAT. The room contained an array of age-appropriate toys and the child was instructed to play with whichever toys he/she chose, and asked either to tell a story or to put on a show for the experimenter with each toy selected. Two independent raters recorded the presence of aggressive behavior from behind an observation mirror. It is not clear from the report whether or not the observers were aware of each child's diagnostic status.

Subsequent to the laboratory assessments, each child's teacher was mailed a copy of the Behavior Problem Checklist (Quay & Peterson, 1967) and was asked to rate each child on the aggression scale. Again, it cannot be ascertained whether the teachers were blind to the child's clinical status.

Reidy (1977) found that the abused children exhibited significantly more fantasy aggression on the TAT than the neglect and control groups. The latter two groups did not differ significantly from one another. An inspection of the means and standard deviations of the three groups revealed a significant amount of overlap, even though one of the overall group comparisons was significant. It would be interesting to examine individual as well as group differences in order to learn about any patterning of competencies that may characterize these maltreated children. Almost all analyses of the developmental sequelae of maltreatment have focused on group differences and on the detection of incompetencies. Very little work has been concerned with the development of competence in these children. This is crucial information to obtain, especially if we are interested in assessing the continuity of adaptation and maladaptation in these children and in identifying stress resistant, "psychologically invulnerable" children (see Garmezy, 1974; Pavenstedt, 1965; Rutter, 1978; E. Werner & Smith, 1982).

In the analyses of the play data, Reidy (1977) found that abused children showed significantly more aggressive behavior than the neglect and the control groups. Interestingly, the children in these latter two groups hardly ever exhibited aggressive play behavior.

Finally, teachers rated abused and neglected children significantly more aggressive on the Behavior Problem Checklist. The abuse and the neglect groups did not differ. However, as was the case for the TAT data, there was a significant degree of overlap among the three samples of children.

When Reidy (1977) analyzed for the effects of residence on aggressive behavior in maltreated children, the only significant result he obtained was that the abused children reared in their homes revealed more fantasy aggression than the abused children residing in foster-care placements. Since many analyses were undertaken, and since this result was not specified as an *a priori* hypothesis, it is likely that it is a spurious finding.

Reidy's (1977) data are interesting, despite some of the possible methodological shortcomings noted, in that they are consistent with earlier clinical reports that abused children display more hostility and aggression in their behavior (e.g., Elmer, 1967; Galdston, 1965, 1971; Johnson & Morse, 1968; Kempe & Kempe, 1978). Moreover, these data illustrate how research in child maltreatment can complement and extend our knowledge about normal developmental theory. In a classic work, Sears, Maccoby, and Levin (1957) found that children who were severely physically punished displayed aggressiveness in their behavior. Furthermore, Hoffman (1960) showed that children who experienced physical and verbal aggression from their mothers likewise utilized these tactics when interacting with their peer group.

These associations between harsh parental disciplinary practices and the subsequent emergence of aggressive behavior in maltreated children are illustrated further in a recent study conducted by Kinard (1980). Kinard set out to test two explicit hypotheses: (a) that physically abused children would have more negative self-concepts than nonabused children; and (b) that physically abused children would express their aggressive impulses in a more aggressive way, either intropunitively or extrapunitively, than nonabused children.

To subject these hypotheses to empirical test, Kinard (1980) studied 30 legally verified physically abused children and compared them with a group of 30 matched comparison children who had never been reported as maltreated to the state protective services agency. The verification of non-maltreatment, at least in the legally reported sense of maltreatment, is an important procedure to employ, especially when maltreated children are matched with control children sharing all other characteristics in common (see our discussion of Elmer's, 1977, study on pages 169–175).

These two groups of children, between the ages of 5 and 12 years, were matched on age, sex, race, birth order, number of children in the family, welfare status (AFDC, non-AFDC, and foster care), family structure (mother alone, mother and male, and foster care), and type of residence (house/ apartment, public housing, and foster home). Forty percent of the children were between 5 and 6, 20% were between 7 and 8. The remainder of the children fell between 9 and 12 years of age. The vast majority were from the lowest social classes. At least one year had elapsed since the initial physical-abuse report was filed with the state agency legally mandated to receive such reports.

The test battery was comprised of three psychological tests: the Piers-Harris Children's Self-Concept Scale, the Rosenzweig Picture Frustration Study, and the Tasks of Emotional Development (TED) test.

The Piers-Harris scale assesses several self-concept dimensions based on the number of a child's affirmative responses from the total set of 80 questions administered: intellectual ability, anxiety, popularity, happiness, physical appearance, and behavior.

The Rosenzweig Picture Frustration Study consists of 24 drawings of potentially frustrating situations in which either a child or an adult in the drawing is shown as saying something. The subject is asked to provide a response of what he/she thinks the child in the picture might say. Each subject's responses are scored for direction of aggression: intropunitive, extrapunitive, and impunitive.

The TED is a projective test that contains objective scoring criteria designed to assess ego functioning in children. The TED consists of 12 pictures, each depicting a particular task that must be mastered in the domain of emotional development. Children are required to tell a story about each photograph, and the examiner asks a series of directed probes that permit each response to be scored adequately. Five dimensions are scored for each picture: (a) *perception of what the picture is about* (correctly vs. incorrectly perceived); (b) *outcome* (successful vs. unsuccessful story outcomes); (c) *affect or feelings attributed to the child in each photograph* (positive or negative); (d) *motivation for the child's actions* (reality-oriented or fear-oriented motivations); and (e) *spontaneity in describing the story.* Five tasks of emotional development considered to be particularly salient for abused children were included in the assessment battery: (a) establishment of trust in people; (b) separation from the mother; (c) self-concept; (d) socialization with the peer group; and (e) aggression. As will be illustrated in a later section, several of these tasks of emotional development are compatible with the organizational perspective for assessing socio-emotional development in maltreated children (Cicchetti & Rizley, 1981; Sroufe, 1979a).

Each child was administered the psychological assessment battery individually at a room in his or her school. Different persons administered and scored the psychological tests. Although the raters were not given any information about the nature of the study, it is not clear whether or not the psychological examiner was blind to each child's clinical status or to the hypotheses to be tested.

Kinard (1980) found that abused children differed significantly from the controls in all five areas of emotional development. Their responses to the Piers-Harris Self-Concept scale revealed no significant group differences from those obtained by the nonmaltreated controls. However, there were seven items that revealed significant differences indicating that the abused children were more likely to have negative self-concepts. Although the magnitude of these differences were small and conceivably could have been chance occurrences, the items appear to be apt descriptors of maltreated children and are congruent with clinical descriptions of abused children. The physically abused children described themselves as being "sad, unpopular, unhappy, disobedient at home, wanting their own way, doing many bad things, and believing their parents expected too much of them" (Kinard, 1980, p. 691). On the TED self-concept task, 30% of the physically abused children, compared to only 7% of the control group, did not see the photographs as symbolizing a child in a position to evaluate him or herself.

The abused children exhibited more extrapunitive aggression than the nonabused children, who, in turn, were more impunitive than the physically abused children on the Rosenzweig Picture Frustration Study. However, in contrast to one of Kinard's hypotheses, the two groups did not differ on measures of intropunitive aggression. Further analyses of the Rosenzweig Picture Frustration Study revealed that abused children differed only in their responses to photographs depicting child-care interaction. On these items, the physically abused children were more extrapunitive, whereas the controls were more impunitive and intropunitive. Analysis of the TED aggression task revealed that the physically abused children gave more reality-oriented reasons for the child's actions, whereas the nonabused controls gave more fear-oriented reasons. These interesting data suggest that all aspects of emotional development are not necessarily impaired in maltreated children: they underscore the possibility that some aspects of these children's functioning may reveal normal competencies.

Finally, analyses of each of the remaining TED tasks—establishment of trust in people, separation from the mother, and socialization with the peer group—revealed significant abused-control group differences. The physically abused children, in contrast to the nonmaltreated controls, gave more reality-orientations for unsuccessful outcomes on the socialization with the peer group task. The nonabused children were more likely to generate fear-oriented answers. On the trust task, the physically abused children were significantly more likely than the controls to have incorrect perceptions of what the child and woman on the picture are doing. Finally, the physically abused children were more likely than nonmaltreated children to provide fear-oriented reasons for why the child will not leave on the TED task of separation from the mother.

Kinard's (1980) study is one of the best designed studies conducted on the socio-emotional sequelae of maltreatment; nonetheless there are several criticisms that can be levied against it. One obvious shortcoming is that all the data are based on self-report or projective measures, the validity of which have often been called into question by personologists (Mischel, 1968). In one sense, Kinard's outcome measures may be construed as trait-oriented personality assessments of physically abused children.

Moreover, since the same measures were given to a small sample of children spanning an eight-year age range, there is no way to analyze the data for any developmental differences. But age- or stage-related differences may be critically important in studies of the socio-emotional sequelae of child maltreatment. Kinard's decision to treat the children in her study as a single age group may have obscured critical differences between maltreated and nonmaltreated children which may exist only at certain ages. For example, the failure to find differences between the abused children and the controls on the measure of intropunitive aggression may be due to the fact that the majority of children in Kinard's study did not yet possess the developmental capacity to turn aggression in on the self. Selman (1980) asserts that most children develop the capacity "to step mentally outside himself or herself and take a self-reflective or second-person perspective on his or her own thoughts or actions" (p. 38) somewhere between the ages of 7 and 12 years. Perhaps such perspective-taking ability is a social-cognitive developmental prerequisite for turning aggression inward.

Kinard's (1980) data, though nonobservational, provide additional support to the normal child developmental literature on parental punishment and aggressive behavior. For example, it is noteworthy that Kinard found that physically abused children differed from nonmaltreated children in their expression of aggression on the Rosenzweig Picture Frustration Study only when the frustrating situations depicted peer-group interactions. One possible interpretation of these data is that the physically abused children, fearful of counteraggression from their parents, displace their aggression on their peer group or other adults (see also George & Main, 1979, 1980). Although a higher incidence of intropunitive aggression in physically abused children was not found in Kinard's experiment, it should be recalled that other investigators have found that physically abused children do turn aggression inward

(Green 1968, 1978b). S. Feshbach (1970), in a comprehensive review of the aggression literature, concluded that the expression of aggression through the fantasy mode is related to the severity of punishment children receive. Likewise, Sears *et al.* (1957) demonstrated that children whose mothers utilized severe punishment practices exhibited a greater amount of fantasy aggression than children whose mothers employed less severe punishment. Again, these data are supported by and compatible with the results of Kinard's (1980) study.

Aragona and Eyberg (1981) complain that the problem of child neglect has received much less scientific attention than the problem of child abuse even though neglect is estimated to be more prevalent and its physical and emotional sequelae are as serious. They note that previous research (Disbrow, Doerr, & Caulfield, 1977; Egeland, 1979; Gaines, Sangrund, Green, & Power, 1978) suggests that interactional measures of mother/child relationships using direct observation methods prove better at differentiating neglect, abuse, and comparison families than personality measures. They cited other researchers who have found similarities in interaction patterns between maltreating families and families with behavior problem children (Burgess & Conger, 1978, 1978; Reid & Taplin, 1977). Drawing on these two insights, Aragona and Eyberg studied three groups of children and their mothers, using an interactional paradigm and an observational method. The three groups consisted of low-income Caucasian mothers and their neglected, behavior-problem and nonproblem children selected from files at the University of Oregon Health Science Center. The results of their study point to the importance of considering alternative pathways to similar developmental outcomes.

Using a reliable and valid system for coding dyadic parent-child interactions (Eyberg, Robinson, Kniskern, & O'Brien, 1978), they found that, compared to controls, the interactions of neglect mothers were most negative and controlling during child-directed play, whereas the interactions of mothers of behavior-problem children were more negative and as controlling as neglect mothers during parent-directed play. Despite these significant differences in the processes of mother/child interaction, the neglect mothers and problem-behavior mothers reported comparable levels of actual behavior problems for their children. Whereas the neglect and the behavior-problem groups significantly differed from the controls, they did not differ from each other. Neglected and behavior-problem children appear to be comparably behaviorally disordered, but they seem to have arrived at this point by different etiological pathways with different implications for prevention and for treatment.

Aragona and Eyberg's specific critique of the use of personality measures to differentiate abuse, neglect, and control families should not be construed as a global criticism of all noninteractional measures for all types of research on child maltreatment. Measures of salient developmental processes have also begun to provide a fruitful alternative to static measures of personality traits in studying the developmental sequelae of child maltreatment. Baharal, Waterman, and Martin (1981) endeavored to explore whether they could identify forms of psychological damage in abused children who did not suffer from neurological trauma, mental retardation, or extreme social and economic deprivation. They chose to focus on the development of social cognitive processes which they considered important to both theory construction on the intergenerational transmission of patterns of child abuse as well as to the differential diagnosis and treatment of psychological harm in maltreated children. Seventeen abused children and 16 controls, all referred by Colorado's Department of Social Services, voluntarily participated in the study. The children were between six and eight years of age; 12 children in each group were boys. The abused and comparison children came from similar lower class family backgrounds. Approximately one-half of each group lived in single-family homes. All the abused children suffered documented physical abuse more than one year before inclusion in the study. It is unclear whether the three experimenters were blind to the children's group status.

The measures of social cognition employed by Baharal and his colleagues included assessments of the following component processes: locus of control (Mischel, Zeiss, & Zeiss, 1974); accuracy of labeling of specific affects and motivations (Rothenberg, 1970); perspective-taking logic (Flavell, Botkin, Fry, Wright, & Jarvis, 1968); social role concepts (Watson, 1977); and moral judgment (Costanzo, Coie, Grumet, & Farnill, 1973). In addition, each child received a test to assess verbal intelligence (Slosson, 1961).

The results of this study are intriguing. Even though the abused children obtained average IQ scores ($M = 102$), they scored significantly lower than the well-matched controls ($M = 112$, $p < .02$). Because IQ often correlated with social-cognitive competencies in one or both groups, the study of group differences on the other measures employed analyses of covariance. Baharal et al. found that the abused children were more likely to perceive outcomes to be determined by external factors and were less effective in understanding complex social roles than were the comparison children, even after IQ effects were controlled for. The abused children also demonstrated diminished abilities to label feelings accurately and to decenter cognitively, though the strength of these findings was reduced when IQ was partialed out. The two groups did not differ in level of moral reasoning.

The framework in which Baharal et al. interpreted their findings is nearly as important as the findings themselves. First, the authors noted the similarity between the descriptions of the social-cognitive styles of these abused children and earlier descriptions of the styles of abusive parents (Newberger, 1979). They concluded that prevention and treatment efforts should focus on enhancing the social perspective-taking skills, social sensitivity, and sense of efficacy of abusive parents *and* abused children if the intergenerational cycle of abuse is to be broken.

Second, although low IQs are often thought to cause diminished competencies in other areas of ego functioning, Baharal and his colleagues suggest an alternative point of view. Perhaps the social-cognitive incompetencies of maltreated children limit their performance on intellectual tests. Children who are especially dependent on external cues in evaluating their performance may suffer in testing situations that require intrinsic motivation and self-feedback. These data strongly suggest the need for longitudinal research to determine whether physically abused children run an increased risk of academic and interpersonal adjustment problems due to their impairments in social-cognitive competencies.

R. C. Herrenkohl and Herrenkohl (1981) have examined the coping strategies employed by both maltreating and nonmaltreating families. These families have preschool children ranging in age from 18 through 71 months. Four groups of families participated in this study. Seventy-two families who had one or more maltreated children and had received less than a year of child welfare services comprised the experimental group. Three control groups served as the basis for comparison: (a) 74 nonmaltreating, protective service families served by the child welfare agency; (b) 50 families served by Head Start who were not receiving any welfare services; and (c) 50 families served by day-care programs, also not receiving child welfare. The protective service group was of slightly higher financial status than the maltreatment families. The Head Start families had the lowest income, whereas the day-care group had the highest income, with approximately one-third of these families earning more than $10,000 annually.

Parent-child interactions were collected in each family's home by two independent raters. The parent and child worked on four tasks selected to elicit achievement orientation, affection, verbal, and nonverbal interaction. It is unclear whether the raters were blind to the diagnostic status of each dyad.

Observations of the four groups of children were made in their preschools. Free play behavior was observed on two half-hour sessions, one week apart. Again, it is not specified whether or not raters were aware of the clinical status of the children. Ratings were made of a variety of behaviors, including aggression or withdrawal reactions to frustrating situations, leadership of peers, withdrawal from peers, and task-oriented behaviors.

The entire set of parent-child interactions has yet to be analyzed, however, preliminary analyses of the 80 mother-child pairs studied reveal that there was less parental approval behavior in the maltreatment group than in the three comparison groups. Moreover, lower level of parental approval of the child was associated with a greater degree of parental detachment from the child as rated by the observers of the parent-child interactions (R. C. Herrenkohl & Herrenkohl, 1981). These findings are congruent with an earlier report by Burgess and Conger (1978) that abusive and neglectful families display more negative and fewer positive interactions with each other than did control families. However, the findings of the Herrenkohls (1981) must be regarded as tentative, not only because all data have yet to be reported, but also because potential lack of rater blindness may render the conclusions tenuous. Moreover, the ages of the mother-child dyads analyzed have not been reported. Thus, no developmental trends can be ascertained from their data.

Based on the analysis of the preschool children, R. C. Herrenkohl and Herrenkohl (1981) conclude that the maltreated children were significantly more aggressive with their peers than were the children of the three comparison groups. Aggressive behaviors were found to occur chiefly in response to difficult tasks or to interfering behavior by peers. These results corroborate and extend the increased incidence of aggressive behavior in maltreated children reported by clinical and experimental studies. The aggressive responses to frustration manifested by the maltreated children suggest that, even at a very early age, they have difficulty resolving conflict, an issue presumed central in the formation of friendship. Since R. C. Herrenkohl and Herrenkohl (1981) have observed a wide age range of maltreated children in preschool settings, it would be interesting if they undertook developmental analyses of their data.

Although most cross-sectional studies of the socio-emotional sequelae of maltreatment have been conducted on school-age children, two studies, one of maltreated toddlers, and one of maltreated-infants, deserve mention.

Relich, Giblin, Starr, and Agronow (1980) observed 13 physically abused toddlers from an urban hospital and compared their motor and social contact behaviors with a matched control group of 13 toddlers drawn from the same hospital who typically were admitted with respiratory ailments. The toddlers, drawn primarily from the lowest social classes, were matched on age, sex, race, and social class of the head of family household. The abused toddlers ranged from 29 months to 53 months of age (mean age = 38 months), whereas the controls ranged from 24 to 47 months (mean age = 37 months). Both groups were free from any obvious neurological impairment and were recruited by social workers from the hospital. The authors do not stipulate how they verified that the experimental group was actually abused.

Relich et al. (1980) videotaped the interactions of the children with their mothers while the dyad was engaged in play or talking to each other face-to-face. The interaction went on for approximately seven minutes. Children and mothers were videotaped in a laboratory setting. One camera was focused on the child from behind the mother, while a second camera focused on the mother from behind a one-way mirror. A split-screen generator was utilized so that both partners were seen on one screen simultaneously. The authors do not comment on the possible effects that the visible camera might have had on the child's behavior.

Six 30-second segments were scored on the seven minutes of dyadic interaction: the first and last minute and the 30 seconds preceding and following the midpoint of the observation period. Children's motor and social behaviors were evaluated using a checklist that assessed six major categories: manipulation, movement, positioning/posture, muscle tone/coordination, mutual exchange, and social contact. It is not specified whether or not the raters were aware of the diagnostic status of the dyads that they viewed.

Matched-pair analyses revealed no statistically significant differences between abuse and control dyads, due chiefly to the large standard deviations obtained. Since some researchers in the field of child maltreatment have followed Sameroff and Chandler's (1975) lead that the quality of care that a child receives is potentially a major factor influencing child development (cf. Martin, 1976; Martin & Beezeley, 1977), Relich, et al. (1980) classified families according to the quality of mother-child interaction. This was accomplished by sorting familes based on their scores on three factors from Caldwell, Heider, and Kaplan's (1966) Inventory of Home Stimulation (HOME). The factors chosen, presumed to best characterize the quality of mother-child interaction, were: (a) emotional and verbal responsivity of the mother; (b) avoidance of restriction and punishment; and (c) maternal involvement with the child. Subgroups of abuse and control children were formed by performing median splits on the home-environment factors.

Statistical analyses revealed few significant differences in the comparisons of the four groups compiled by the categories of abuse and control and two of the three factors based on the HOME classification. However, one of the HOME factors, avoidance of restriction and punishment, did reveal significant group differences.

Three major findings emerged from the *post hoc* comparisons employed by Relich and her colleagues. First, the most disadvantaged abuse group (i.e., the less favored on the avoidance of restriction and punishment factor on the HOME) exhibited a greater amount of self-manipulation than the three other groups. Relich *et al.* (1980) interpret these findings to reflect the fact that maltreated children must resort to more covert, body-movement forms of communication, presumably because they do not possess the trust necessary to express their emotions openly. The favored abuse group showed the highest level of object manipulation. Second, the less-favored control group exhibited more negative affect than the other three groups. Relich *et al.* (1980) propose that the absence of abuse fosters the appearance of more overt forms of feelings and emotions, especially in conjunction with a restrictive and punitive environment. Third, the less-favored control children displayed significantly fewer instances of object manipulation and significantly more negative affect than the favored abused children. Thus, physical abuse *per se* does not appear to affect inexorably the openness with which abused children express their emotions. Although these results are intriguing, the small sample size employed and the absence of analyses for developmental differences across the two-year age range studied suggest that these tentative findings be further investigated.

Gordon (1979) examined the nature of infant-mother attachment in a group of infants diagnosed as "nonorganic failure to thrive (FTT)." These infants were hospitalized in the first year of life for delayed or arrested physical growth (below the third percentile for height and weight) that could not be traced to organic illness. Their social behavior was also of concern to hospital staff who described them as "apprehensive," "frightened," "apathetic," and "withdrawn." Various authors have sought to find the cause for the nonorganic FTT syndrome in the mother-child relationship. This tendency is reflected in such descriptions of the syndrome as "deprivation dwarfism" (Silver & Finkelstein, 1967), "maternal deprivation syndrome" (Patton & Gardner, 1962), and "masked deprivation syndrome" (Prugh & Harlow, 1962). However, these studies tend to view the infant as a passive victim of the mother's caregiving and ignore the question of the infant's contribution to its own vulnerability.

Employing a laboratory paradigm similar to Ainsworth's "strange situation," Gordon observed 20 nonorganic FTT infants aged 12 to 19 months and 20 controls, matched on age, race, sex, social class, length of hospitalization, and age at hospitalization. The control group of infants had been hospitalized due to accidents they had incurred. Infants' attachments were classified based on Ainsworth's classificatory system (Ainsworth, Blehar, Waters, & Wall), 1978), as securely or insecurely attached. Fifty percent of the nonorganic FTT infants were classified as insecurely attached as compared with 15% of the controls. In addition, three variables were found to be associated with insecure attachment: mother's depression, infant temperament, and low socioeconomic status (SES). Infants whose mothers were rated as depressed were more than three times as likely as other infants to be insecurely attached (64% vs. 21%); infants rated as low on the persistent dimension of temperament were more than twice as likely as other infants to be insecurely attached (53% vs. 20%); and infants from families in the lowest social class were more than three times as likely as other infants to be insecurely attached (64% vs. 21%). One subgroup of the nonorganic FTT infants was of particular interest. Five of these infants were rated as low on the persistence dimension of temperament *and* had mothers rated as depressed. Each of these five infants was insecurely attached.

Gordon's study demonstrates both the vulnerability and resiliency of children in high-risk groups. A significantly higher rate of insecure attachment was found in the nonorganic FTT infants than in their controls, indicating the serious developmental risk of these infants. At the same time, half of the nonorganic FTT infants were able to develop secure attachment relationships. Clearly there are multiple pathways to secure and insecure attachment.

Gordon stresses the need to understand the processes by which characteristics of infants and their mothers are transformed into attachment relationships. For example, the depressed mother and her temperamentally atypical infant interact to produce what Stern (1977) calls a "regulatory failure" in which the mother is not "adequately stimulated by the baby to produce the behaviors that will stimulate her to stimulate him . . . and so on." As a result the infant learns that his or her affective expressions are ineffective in eliciting responses from the mother. The infant then approaches the task of attachment without the basic trust (Erikson, 1950) in his or her own ability to explore the environment and his or her mother's availability as a "secure base" (Ainsworth, 1973).

2.3. Follow-up Studies of the Developmental Sequelae of Child Maltreatment

Several investigators have conducted follow-up studies of maltreated children. Typically, the children were evaluated once in the early years of life and then subsequently reassessed a number of years later. These studies, though far from being methodologically flawless, represent important advances over cross-sectional studies in that they enable us to begin to document the longer term sequelae of abuse and neglect.

Morse *et al.* (1970) conducted a follow-up study of 21 children who had been hospitalized approximately three years earlier for injuries judged to be sequelae of "suspected" child abuse or gross neglect. Only one confirmed episode of maltreatment was required for inclusion into this study. Thus, unlike some reports in which only severely maltreated children were studied (e.g., Elmer, 1967), these children had not all been chronically and repeatedly maltreated. The follow-up period ranged from 2 to 4½ years after the first incident; the ages of the children spanned from 33 months to 10 years (the median age at follow-up was 63 months). No control group of children participated in this follow-up investigation.

Data on the emotional development of these "suspected" abused and neglected children were gathered at interviews and home observations. Children living at home were observed in that context, while their parents were interviewed about their children's social and emotional development and any social or emotional problems. If a child was residing with foster parents, then he or she was seen in that setting while the foster mother was interviewed. One person, who was aware only of the medical findings at the time abuse and neglect were suspected at the initial evaluation (Holter & Friedman, 1968a), and not of the alleged circumstances surrounding the report, conducted the interviews. In addition, this interviewer recorded her impressions of the child based on her direct observations of the child and the mother-child relationship.

Based on the interview and observational data, only 6 of the 21 children (29%) were judged to be within normal limits cognitively and emotionally at the time of follow-up. Of the remaining 15 children, 9 were judged to be mentally retarded and 6 judged to be emotionally disturbed. For the 6 children who were developing normally, the defining characteristic appeared to be that their mothers *perceived* their parent-child relationship as good. These 6 mothers described their children as affectionate, easy to discipline, and problem free. In contradistinction, the mothers of the 7 children seen as most grossly disturbed all reported poor mother-child relationships. As Sameroff (1975) has underscored, parental perceptions of the child may exert a crucial influence on the child's development (see also Friedrich & Boriskin, 1976; and E. C. Herrenkohl & Herrenkohl, 1979; R. C. Herrenkohl & Herrenkohl, 1981). The 9 mentally retarded were perceived very adversely by their parents. Three were considered to be "sickly," while the remaining 6 were seen as "bad," "selfish," "spoiled rotten," and "defiant."

Six of these 21 abused/neglected children were found to have experienced additional maltreatment after the initial report. Five of these children were judged to be either emotionally disturbed or mentally retarded.

The high percentage of emotional problems, as well as the other developmental characteristics of this sample, is compelling. However, the lack of interrater reliability and systematic observations of the children, combined with the absence of a control group, make the data difficult to interpret.

Friedman and Morse (1974) followed up 41 children from an original study involving 156 cases of injury found in children under 6 years of age in the emergency department of a university medical center (Holter & Friedman, 1968b). Fifteen "suspected" abused, 7 "suspected" neglected, and 27 randomly chosen "repeated accident" children were studied, five years after the emergency room observation. The 41 children ranged in age from 63 to 126 months at the time of the follow-up investigation, with a mean age of

7½, 7, and 8 years for the "abuse," "neglect," and "accident" groups, respectively. No demographic characteristics of the three groups are presented.

Data were collected by a single interviewer, who was not cognizant of whether the initial injury had been judged to fall within the abuse, neglect, or accident categories. However, the interviewer was aware of the date and nature of the injury for each family studied. Furthermore, mother-child relationships were assessed by this same interviewer based on observations of mother-child interaction at the time of the interview in the child's home, and from verbal reports given by the mother about the child's behavior. The areas of child behavior explored with the mother included temperament, obedience, physical activity, sibling and peer relationships, and specific bedtime and mealtime problems.

Freidman and Morse (1974) report that more "accident" children displayed good relationships with their mothers than did those in the other two groups. However, these differences did not reach statistical significance, partly because of the small sample sizes employed. The absence of additional raters, and the unsystematic nature of the observational procedures, make these findings of questionable validity.

Kent's (1976) follow-up study had two main goals: to provide data about the effects of foster-care placement on the later development of abused children; and to provide data on the effects of specific kinds of abuse that may be independent of socioeconomic status (SES). To pursue these goals, he developed a strategy that employed three groups: two abuse groups and a nonabuse control group. The two abuse groups were drawn from children under court dependency in Los Angeles County in November, 1971. One group consisted of 219 inflicted injury or "nonaccidental trauma" (NAT) cases; the other of 159 "gross neglect" (NEG) cases. (Any children who were adjudicated for both reasons were dropped from the study.) One hundred eighty-five nonabuse control group cases were selected from the Department of Public Social Services (DPSS) "specialized protective services" caseload. Although none of the protective services (PS) cases had been adjudicated as abused, this group was selected because they shared low SES and high-risk environments with the two abuse groups. The NAT and PS families differed only slightly in demographic features (PS families reported slightly higher rates of both per capita income and welfare dependency than NAT families). NEG families fell far below both other groups in parent education, parent employment, and per capita income.

The data for the study consisted of the responses of the children's social workers to a precoded questionnaire. To answer the questionnaire, each social worker was requested to consult their case records which included DPSS contact records, hospital and police records, and reports from the children's foster home and school.

The results reported by Kent focused on the differences between the intake and follow-up status of the children in three general areas: behavioral characteristics, developmental characteristics, and school performance. (No data were presented on the length of time elapsed between intake and follow-up, but to be included in an abuse group, a child had to be under court dependency for at least six months.)

On intake, NAT and NEG children posted higher rates of problem behaviors (excessive disobedience, severe tantrums, emotional withdrawal, aggression toward adults and peers, and poor peer relations) than the PS children. In addition, NAT children were rated as more aggressive, more disobedient, and less well related to their peers than NEG children. Most notably, both the NAT and NEG groups improved on nearly all behavior problems from intake to follow-up. The NAT children showed marked improvement in their management of aggression, and both groups improved most dramatically in emotional withdrawal.

The developmental characteristics rated by social workers in Kent's study were delay in motor development, delay in language development, and delay in activities in daily living (feeding, dressing, hygiene, and toileting). The children from the NEG group showed the greatest delay in all three areas on intake, and although improving substantially by follow-up, still remained delayed compared to the NAT children. Corroborating the findings of Elmer (1967) and Martin et al. (1974), both the NAT and NEG groups revealed greater developmental delays on follow-up in language than in the other two domains.

The most dramatic gains posted by the NAT and NEG children were in the realm of school performance. Whereas a majority of NAT (53%) and NEG (82%) children were rated as performing below average or failing in academics on intake, only a minority were so rated on follow-up (28% for both groups). The percentage of children rated as unsatisfactory in peer relationships dropped from 67% for NAT children and 60% for NEG children on intake to 38% and 35%, respectively, on follow-up.

One of the most interesting aspects of Kent's study concerns those children who were "too young" to be rated on behavioral, developmental, and school characteristics on intake but were included in the follow-up. Whereas Kent provides no precise documentation of the children's ages, from an annotation of a table in the article it appears they were under 3 or 4 years at the time of intake. The incidence of both behavior problems and developmental delays in the younger children at follow-up was substantially less than for the older children at intake and follow-up.

The Kent study marks an important advance in the study of the developmental sequelae of child maltreatment. The study makes three important assumptions that dramatically improved the quality of the research. Kent assumes that maltreated children, like all children, function in multiple domains (behavioral, language, motor, and social), change over time (intake vs. follow-up), and are affected differently by different environments (home vs. foster care) and different social histories (nonaccidental trauma vs. gross neglect vs. high risk environment). Its major findings have important clinical and policy implications. The differences among NAT, NEG, and PS children on intake in problem behaviors, developmental delays, and school performance constitute preliminary data on the validity of NAT and NEG as diagnostic classifications (see Aber & Zigler, 1981, for a discussion of reliability and validity issues in schemes for classifying types of child maltreatment). Similarly, the differences between abused children on intake and abused children on follow-up seem to point to the positive impact of intervention, here defined as foster care, on the development of abused children. Finally, the lower rates of problem behaviors and developmental delays in the young children compared to the older children point to the potential benefits of early intervention.

Unfortunately, all of Kent's findings, as interesting as they are, must be considered equivocal due to major limits in the scientific design of the study. First, the data on the validity of the diagnostic groups must be suspected because the social workers who rated the children were not blind to the reasons for adjudication. Second, the data on the benefits of intervention are also questionable on the grounds that intake and follow-up ratings were neither blind nor independent. In addition, without the use of nonintervention abused comparison groups, one cannot rule out the possibility that what appear to be intervention effects are instead maturational effects. Kent's design inextricably confounds treatment and aging effects. Finally, any firm conclusions about the beneficial impact of early intervention must await much more thorough and detailed age/stage analyses. Despite these methodological flaws, the Kent effort should be considered a valuable precursor to studies that tackle (a) the problems of the reliability and validity of classification schemes, (b) the need for multivariate outcome data, and (c) the need for longitudinal research designs.

The most recent follow-up study of the developmental sequelae of child maltreatment is also the most important and the most controversial. Previously, none of the follow-up studies employed a matched comparison group. But Elmer (1977) studied abused children matched on age, race, sex, and SES to a group of accident children and a nontrauma comparison group. Most children in all three groups, consequently, were from the lower classes (defined as classes IV and V on the Hollingshead Two-Factor Index of Social Position). The abused children and accident children had been studied in Elmer's original paper (1967) when they were 12 months or less. The follow-up study presents data on 17 abused children and 17 accident children, eight years later. In addition, Elmer wished to match on history of hospitalization during infancy and so assembled a nontrauma comparison group of 25 children. (Nine pairs of abused/accident children were similar as to hospitalization during infancy; 8 pairs were dissimilar. For the 8 dissimilar pairs, 2 comparison children were needed, one resembling the abused child's hospitalization history, one resembling the accident child's history. For the 9 similar pairs, only one comparison child was needed, $8 \times 2 + 9 = 25$.

Data for the study were collected in three settings. First, mothers were interviewed at home by women interviewers of the same race about their perceptions of their child and their methods of reward and punishment. Second, all children were brought to Elmer's labs for half-day evaluations by pediatricians, psychologists, speech pathologists, psychiatrists, and social workers. All clinicians performing the evaluations were blind to each child's group status. The evaluations covered a wide variety of areas of development and assessment techniques including the child's health and physical development (assessed through pediatric exam and anthropometric measures), intellectual functioning (by analysis of a child's entire school record, including tests administered, grades, and grade placement), language development (articulation ratings through the child's response to pictures on a projective test), self-concept (the Piers-Harris Self-Concept Scale), impulsive and aggressive fantasy material (through a dramatic puppet play situation) and the child's control of aggression and degree of disturbance (through clinical ratings). Finally, teacher ratings of the child's school behavior and attitudes were collected (by precoded questionnaire).

The major finding of the study was the absence of group differences in health history, intellectual status in school, language development, self-concept, and the clinical ratings of control of aggression and degree of disturbance. Slight group differences were found in some anthropometric measures (paradoxically, abused children weighed significantly more than nonabused comparisons) and measures of impulsive and aggressive fantasy material (the abused children's role-play creations became increasingly impulsive and aggressive over a series of 5–6 stories). But these differences were found between the abused children and either the accident *or* the comparison children, not both groups. Elmer interprets this fact as tending to weaken the results. Curiously, the largest group differences were found between two subgroups of abused children on the language measures where children in their own home performed better than children in foster care. Elmer correctly points to the potential influence of class and race on these findings. Six of eight abused children in foster care were lower class black children; six of nine children in their own homes were white children, four of these from the middle class.

Elmer entertains four possible explanations for the lack of systematic differences between abused children and matched comparisons. She discounts the notions that children from Pittsburgh (the site of the study) or children from hospital populations experience poorer child-rearing methods than peers in other communities or from other referral sources. She also states that she posesses no conclusive data that the entire sample of children, not just the abused children, experience repeated aggression from their lower class caretakers, this despite considerable anecdotal evidence of family chaos and disorganization, parental alcoholism and drug abuse, and multiple examples of intrafamilial violence (including spouse/spouse abuse and parent/sibling abuse) presented for all three groups. She instead opts for the notion that lower class family status may exert as much influence on later child development as does a history of abuse.

Elmer's findings and interpretations raise a host of complex scientific issues with policy implications. The first set of issues concerns flaws in the measurement and conceptualization of the two predictor variables "abuse versus accident" and "social class." As previously described, Elmer used as her measure of social class the Hollingshead Two Factor Index of Social Position (Hollingshead, 1956). In a recent review of how social class is measured and conceptualized in child development research (Mueller & Parcel, 1981), this method is criticized as too crude to capture the true variance in social status among child-rearing families. Rather than Hollingshead's 6-point scale, the authors suggest the adoption of occupational prestige scores as a more valid and refined measure of those dimensions of social class most relevant to child development research: the ability to command various forms of social, cultural, and economic resources (see Mueller & Parcel, 1981, for a more elaborate description of occupational prestige scores). Occupational prestige scores offer four other major advantages over Hollingshead social position scores: they have been normed on excellent national samples, they systematically capture more variability in social status (they are expressed on a 0–100 scale), they have led to derivations of prestige scores for "housewife" and "looking-for-work" categories, and they have led to computed scores for family prestige that can accommodate important variations in family structure (e.g., single-female head-of-household vs. two-parent, one-wage earner families vs. dual-career families). These advantages become even more important when one wishes to study the impact of lower socioeconomic status on child development. The prevalence of temporary unemployment and female-headed households, and important distinctions among various strata of the working poor should no longer be ignored in developmental research, especially in such a policy-sensitive and important field as child maltreatment.

The conceptualization and measurement of the "abuse versus accident" distinction constitutes another flaw of the Elmer study. The weakness does not lie principally in the initial assessment of "abuse" when the children were infants. In 1967, Elmer employed a method of classifying a child as abused that was defensible by scientific standards of the time.

> A child was judged abused if one or more of the following criteria were present: report or admission of abuse of the patient or a sibling; injuries incurred more than once; or history conflicting or inadequate to explain the child's condition. (Elmer, 1977, p. 274)

The judgment was made jointly by a pediatrician, a social worker, and the author on the basis of a pediatric and developmental assessment in the outpatient department of the hospital and a maternal interview in the home about circumstances of the index event. All three judges had to agree that the child was abused; *otherwise the child was considered an accident victim.* Despite the vagaries of the individual criteria, Elmer's decision to require unanimity among the judges probably served to drastically reduce the number of false positives in the abuse classification.

The major weakness of the method lies in the fuzzy distinctions between "abuse" and "accident" groups. By reassigning cases to the accident group, unless the judgments of abuse were unanimous, Elmer probably inadvertently increased the number of false positives in the accident classification. No definitive criteria for inclusion into the accident class are presented. Thus, it is reasonable to assume that the accident class included some abused children who were not unanimously judged as abused. A sounder strategy would have excluded from the accident classification any of the children judged by even a single evaluation as abused. In short, the group assignment method virtually assured substantial overlap between the two groups.

A second source of confusion between the abuse and accident groups that could serve to reduce group differences is implicit in the study. On the basis of developmental studies by Piaget and his colleagues (Piaget & Inhelder, 1969), we know that the first year of life is a time when the infant lives without any capacity for evocative recall (imaging of previously experienced situations). The capacity for the independence of memory from the trigger of immediate experience does not develop until around the middle of the second year of life. This means that all the infants diagnosed either as accidents or as abused by Elmer did not have the capacity to recreate a former experience when they were subjected to the index event of abuse or accident. Instead, they could only recognize the abuse or accident if it occurred again. Similarly, the infants did not perceive their environments in terms of figure and background differentiations that allow the construction of animate and inanimate object configurations. For these reasons, although the distinction between the trauma caused by abuse and by accident has enormous significance for adults, it is likely to have little or no significance for infants under 12 months. From the infants' point of view, an accident and an abuse incident may bear remarkable similarity to one another. The Elmer study retains an unrefined notion of early vulnerability not in keeping with basic scientific findings of developmental psychology.

A third factor that could obscure group differences was alluded to by Elmer but then dismissed: the possibility that children in the accident and comparison groups were also abused by their lower class caretakers. Although Elmer inquired into the protective services histories of accident and comparison group children and excluded any children from these groups with a positive history of abuse, the possibility remains that some children suffered undetected incidents of abuse. An even stronger possibility is that the accident and comparison group children experienced other undetected forms of maltreatment (e.g., emotional abuse) (see Egeland & Sroufe, 1981) that frequently accompany physical abuse (Giovannoni & Becerra, 1979). In that case, the lack of group differences could be understood as reflecting the influence of undetected forms of maltreatment common to all three groups (and different than lower class status). Firm conclusions about the differential effects of lower class status and maltreatment on child development must await the development of research strategies that independently assess childrearing histories, quality of parenting and of home environments, and maltreatment histories of comparison children as well as children suspected of being maltreated. If basic dimensions converge to paint a consistent picture of the quality of the parent/child relationship, the hypothesis that children may be suffering undetected forms of child maltreatment can be ruled out with more confidence.

Two other aspects of Elmer's study that may serve to mask group differences involve the choice of a matching strategy and the selection of outcome measures. Although the importance of comparison groups in the study of the consequences of child maltreatment is now undebatable, the choice of an overly rigid/inclusive matching strategy at this stage of research is debatable. If members of comparison groups are matched to maltreatment group members on too many characteristics, this may serve to skew how representative the lower class comparisons are of the population of lower class children.

The logic of the construction of the comparison groups in studies of the sequelae of maltreatment should conform to the precise nature of the questions asked and should take into account the populations to which one wishes to generalize the findings. In Elmer's study, the decision to match the abuse and accident groups with the comparison group on both SES *and* history of hospitalization during infancy may have served to skew severely the representativeness of the comparison group. It would be very important to evaluate the rate of infant hospitalizations in the population of lower class families before one generalized the findings from this particular lower class sample to all lower class families. In other words, Elmer's conclusion that low SES may have as profound an impact as abuse on child development rests on the representativeness of her lower class comparison group. A stronger, sturdier inference would be that those sociodemographic features of a family that lead to infant hospitalization may have as strong an impact on child development as the impact of a history of maltreatment. For a discussion of the distinction between the etiology of child maltreatment and pediatric social illnesses see Newberger, Reed, Daniel, Hyde, & Kotelchuck (1977).

Finally, Elmer's study points to the importance of the selection of proper outcome measures in the study of the sequelae of child maltreatment. The selection of measures should be guided by notions of stage-specific developmental tasks on which the child may prove competent or impaired. The cutting edge of development is the sum of new competencies a child is attempting to master. Performance on age-salient new tasks may be more likely to reveal differences among children of various socialization histories than age-inappropriate or non-age-specific tasks. By the early school age years, the social-emotional and motivational subsystems of development may be much more open to modification by different experiences than are the more closed, genetically determined physical, cognitive, and language subsystems of development. Thus, if one were searching for effects in a single, isolated area of development, one is more likely to discern group differences in the socio-emotional and motivational domain than in the physical, cognitive, and language domains. However, researchers are not restricted to consideration of one domain at a time. The most intriguing and fruitful advances of research in development presently include examination of the interrelationship among various subsystems of development (Cicchetti & Pogge-Hesse, 1981; Cicchetti & Schneider-Rosen, in press; Cicchetti & Sroufe, 1978). A developmental approach to the sequelae of child maltreatment would rely less heavily on static or product measures of developmental status or personality traits and more on fluid measures of the component processes of children's performance on various age-appropriate tasks.

In view of these remarks concerning the selection of outcome measures, Elmer can be criticized for failure to identify stage-salient developmental competencies of assessment; overemphasis on physical, language, and cognitive development rather than socio-emotional and motivational development; failure to treat outcome as a multivariate concept; and overreliance on product measures. The potential of a more process-oriented approach to assessment of socio-emotional development is indicated by the results on

Elmers' only process measure, the tendency for maltreated children to become increasingly impulsive and aggressive in their role play the longer they endured provocative stimuli.

Elmer and her colleagues were struck by the pervasiveness and degree of the psychological handicap of the children in all three groups. Her conclusion that social class is as powerful an influence as abuse on later child development places the issue of what constitutes "developmental risk" in a stark sociopolitical light. Here, we are not disagreeing with her conclusion so much as suggesting that the findings of this study may be considered tentative until some of its scientific flaws are transcended by other investigators and its findings are corroborated or challenged.

2.4. Theoretically Derived Studies of the Socio-emotional Sequelae of Child Maltreatment

One striking feature of the cross-sectional and follow-up studies is their collective failure in specifying stage-specific developmental tasks or age-salient developmental competencies that may be at risk due to a history of child maltreatment. Two cross-sectional studies that have assessed competencies of maltreated children are the works of George and Main (1979, 1980) and of Gaensbauer and his colleagues (Gaensbauer & Sands, 1979; Gaensbauer, Mrazek, & Harmon, 1980).

Although they approach their work from different professional vantage points—George and Main from academic developmental psychology, Gaensbauer and his colleagues from clinical pediatrics and psychiatry—they draw upon similar bodies of previous theoretical and empirical work on the normal development of the attachment and affective communications' systems between caretaker and child to study the development of maltreated children of roughly the same age (6–36 months).

Probably the most detailed study of the behavioral consequences of abuse to date is that of George and Main (1979, 1980). These investigators were particularly interested in the consequences of physical abuse with respect to the organization of attachment behavior. They reasoned that the child would most likely experience physical abuse as a rejection by the caregiver. Main's (1977, 1980) previous research with *normal, nonabused* children had shown that rejection by the caregiver is related to avoidance of the mother in stressful circumstances; to the infant's failure to approach *other* friendly adults; and to angry behavior directed toward the mother (hitting or threatening to hit the mother, active disobedience of commands by her). George and Main suspected that abused children would show extremes of the classes of responses that characterized the rejected (but nonabused), infants.

George and Main observed 10 infants/toddlers between the ages of 12 and 36 months in day-care centers that had been especially designed to be therapeutic for physical abuse cases. They watched the children during several types of activity over a three-month period. They paid especially close attention to the socially directed behavior of the children looking for instances of *approaches* to others (adults and same age peers), *avoidance* of others, and *aggressive* instances. A fourth category was established that concerned what they termed *approach-avoidance*. This last category was used to characterize instances in which the child showed both approach and avoidance behaviors at the same time (e.g., crawls toward the caregiver with head and eyes turned away), or in the same behavioral sequence (e.g., infant creeps toward another child but suddenly veers away before making contact).

In order to determine whether or not the patterns of behavior observed in the group of abused children differed from what might be expected of other children who had not been abused, George and Main compared the frequencies observed to those of a control group of 10 children. These control children were drawn from the same socioeconomic population, and were matched for age, mothers' education levels, and for marital status. Unfortunately, the peer observations could not be made blindly, since the physical characteristics of the physically abused children made their clinical status obvious. However, three of the four observers were unaware of the experimental hypotheses.

George and Main (1979, 1980) found many differences between how abused and nonabused children interacted with both peers and caregivers. As a group, abused children were less likely to approach the adult caregivers and were more likely to show avoidance to approaches made by both adult caregivers *and* their agemates, to direct aggressive responses to adults, and to respond to a friendly approach on the part of both peers and adults with approach-avoidance behaviors.

Within these behavioral categories, there were some other differences in the ways abused children's behavior was organized when compared to the nonabused sample. With respect to approaches made to the caregiver, George and Main note that both abused and nonabused children made occasional *spontaneous* approaches to the adult, and that the differences in this aspect of approach behavior do not discriminate between them. However, in response to a *friendly* overture by the adult, the abused children failed to approach as often as did the nonabused children. Further, when the abused group did approach the adult in response to a friendly overture, they were likely to approach indirectly. That is, they would come to the adult from the side or from behind, or, rarely, by turning around and walking *backward* to the adult. There were also interesting differences with regard to aggressive behaviors that discriminated the abused from the nonabused group. There was a difference in the amount of aggression directed to caregivers. Specifically, abused children were more likely to physically assault or threaten to assault caregivers than were the nonabused children. Further, abused infants physically assaulted other infants nearly twice as often as did the control-group infants.

In discussing their results, George and Main note that the majority of the differences between abused and nonabused infants in their study concern the organization of behavior with respect to caregivers. They also point out that avoidance and approach-avoidance were most often seen in response to friendly overtures made by the caregivers. This is anomalous since one might expect that a child, even one who might have reason to be apprehensive of adults, would not be hesitant to approach an adult who was being overtly affiliative. George and Main interpret this intriguing finding by suggesting that avoidance in this circumstance serves the function of enabling the infant to *maintain control* over behavior.

In supporting their arguments that avoidance may be a self-control strategy, they appeal to a wealth of data on animal behavior (see Chance, 1962), in which a similar phenomenon is observed. Many animals find the close proximity of another animal to be a threatening experience that tends to result in "fight or flight" responses if the experience of threat is not somehow modulated. In animals, this modulation is often accomplished by turning the eyes and head away from the opponent (or *friend* in many circumstances). The shift of attention away from the partner, then, permits the maintenance of proximity and the maintenance of self-control.

This model is appealing for understanding the behavior of the abused infants in this study. We may assume that interaction and mutual attention has sometimes led to unpleasant, or even dangerous consequences for the abused infants and that such consequences are likely to have aroused distress, fear, and anger. Therefore, mutual attention is especially threatening to the ability to maintain self-control in these children. If the approach to the adult is spontaneous, the child is in control of the situation; however, in the likelihood of direct attention from the caregiver, avoidance of contact can serve to maintain control. Finally, should the infant choose to approach when contacted by the adult, some degree of arousal can be controlled by approaching in such a way that the likelihood of mutual attention is limited (i.e., by approaching from the side, or rear, with the head turned away). Interestingly, only two of the abused infants were ever observed to turn directly to look at the caregivers who were making a friendly overture, whereas this behavior was seen in all nonabused children.

In concluding this discussion, George and Main note that the behavior of abused children does resemble the behavior of rejected infants from normal samples. They differ from matched controls in that they are more aggressive and they also respond negatively to friendly overtures. These children are interpersonally "difficult."

The data reported by George and Main clearly illustrate the notion of distortion in the developmental agenda. In the normal course of social development, the establishment of an attachment relationship with the parent promoted later *engagement* of the wider social world. The infant whose attachment relationship has been satisfactory is able to enter into the wider social world of peer contacts and of contacts with other adults with the expectation of positive consequences. The abused child, whose experiences with caregivers have not laid the groundwork for positive expectations about social interchanges, tends to behave in a manner that all but forecloses the possibility of successful adaptation in the wider environment. Although these children have, in a sense, adapted to the reality of their living situations, the nature of these adaptations is to prevent, rather than to promote, interpersonal contacts and interactions. This is a developmental distortion, with personal and social (societal) consequences that transcend the infancy period.

Gaensbauer and his colleagues also approached their work on the socio-emotional development of maltreated children from previous studies on normal socio-emotional development, specifically on affective communication between infants and caretakers (Emde, Gaensbauer, & Harmon, 1976; Emde Katz, & Thorpe, 1978: Emde, Kligman, Reich, & Wade, 1978). In their first paper (Gaensbauer & Sands, 1979), their sample consisted of 48 infants between the ages of 6 and 36 months primarily from lower SES families who had been referred by Denver's Department of Social Services for clinical evaluation of suspected or documented abuse or neglect. Two-thirds of the children were evaluated with their natural mothers and one-third with their foster mothers. Besides suspected physical or emotional abuse or neglect, the authors state that the majority of children also experienced problems like chaotic homes, separations from parents, foster placements, and inconsistent caretaking experiences. The research method consisted of the caretaker/infant pair following the same laboratory procedures as used for the normative study of emotional expression in infancy. The session consisted of: free play between infant and caretaker, a stranger approach and pick up, a maternal approach and pickup, a Bayley developmental test with several mild frustrations, and verbal prohibitions and a brief maternal separation and reunion sequence. After the experimental session, the caretaker was interviewed about the similarity of the infant's behavior in the lab and at home, as well as the caretakers' reactions to child behaviors. Although the entire session was videotaped for future analyses, the data for both papers were based on consensual judgments of the authors who, after reviewing the tapes, compared the responses of the maltreated infants with the responses of normal infants observed in their laboratory.

Six varieties of distorted affective communications *from infant to caretaker* were identified and described: affective withdrawal, lack of pleasure, inconsistency/unpredictability, shallowness, ambivalence/ambiguity, and negative affect communications (distress, anger, sadness). Each distorted communication was discussed by contrasting it to affective communications between nonmaltreated children and caretakers, especially emphasizing the impact of distorted communications on the caretakers and on the process of creating successful infant/caretaker partnerships. Gaensbauer conceptualizes these distorted affective communications as the child's contribution to perpetuating (if not initiating) various patterns of maltreatment.

The second study published by this group (Gaensbauer *et al.*, 1980) appears to be a report on a subset of the same sample. The two major changes from the first to the second report are: (a) a reduction in the number of children studied (from 48 to 30) and (b) a different presentation of the outcome variables. Instead of describing six forms of distorted affective communications exhibited in various combinations by most maltreated children, the latter paper attempts to describe "four relatively consistent affective patterns." The authors assume that, "after a careful consideration of the predominant picture," each child can and should be classified into one and only one of these groupings because the groupings "represent crystallizations of what we believe to be the essential elements of the developmental disturbances present in each of these infants." They also present percentage frequencies of each problem in their sample as well as speculate on experiences that may lead to each particular pattern.

The largest group of infants (approximately 40%) were classified as *developmentally and affectively retarded.* The infants in this group were described as emotionally blunted, socially unresponsive, inattentive to their environment, and retarded cognitively, emotionally, and motorically. Gaensbauer speculates that a history of marked deficiencies in caretaker/infant interactions verging on stimulus deprivation has led to this pattern that he likens to Spitz's (1965) "hospitalism" syndrome and describes as the most devastating form of neglect.

Closely resembling the retarded infants are those infants classified as *depressed* (approximately 20%). Although they exhibited some motor retardation, inhibition, withdrawal, and an aimless quality to their play, they were distinguished from the retarded children by the typical facial patterns for sadness and depression, and, most importantly, by their ability to engage and to perform at age level if they were greatly encouraged. Gaensbauer speculates that these infants experienced an initial period of adequate parenting but then, due to abandonment, separation, or maternal depression, experienced a relative loss in the quality or quantity of parenting.

Although all the maltreated infants displayed some degree of affective ambivalence, some infants displayed this tendency so markedly that they were classified as *ambivalent/affectively labile* (25%). They were distinguished from the first two groups because they appeared neither depressed nor deprived, because they performed at age-level on the Bayley, and because they were able to show pleasure and engagement in interactions with their caretakers. Their most characteristic feature was the rapid shifts they demonstrated from engagement and pleasure to withdrawal and anger. Other researchers (George & Main, 1980) describe a similar group of children in behavioral terms as experiencing severe approach/avoidance conflict. Periods of consistent caretaking alternating with periods of maltreatment are thought to be the precursors to their pattern of emotional expression.

Finally, infants who appeared very active and disorganized in their play and exhibited low frustration tolerance and frequent outbursts of anger and destructive behavior were classified as *angry* (15%). Gaensbauer believes these infants come from highly charged home environments where they experience frequent harsh punishments at the hands of their caretakers.

The work of Gaensbauer and his colleagues suffers from some serious limitations; it also possesses unique strengths that in the final analysis outweigh the limits. Although consensus judgments of videotape data mark a significant improvement over single-rater clinical observations, they are still subject to unreliability and other forms of bias. Establishing various types of distorted affective communications and unique patterns of emotional expression that distinguish maltreated infants from nonmaltreated infants is so important clinically, legally, and theoretically (Aber & Zigler, 1981), that these types and patterns should be operationally defined. Then they can be clearly communicated to other investigators and can be tested using independent raters blind to the group status of the infants.

Gaensbauer's approach to studying the socio-emotional development of maltreated children from a broader perspective on normal development led to the fruitful focus on a salient stage-specific competency (affective communications from infant to caretaker) and on other issues of major import to developmental theory in general (for instance, continuity of adaptive style or processes that contribute to the creation of successful parent/infant partnerships). However, without more elaborate and precise data on family demographic characteristics and on child-care practices for both the maltreatment sample as well as the comparison sample, Gaensbauer and his colleagues miss the important opportunity to contribute to the debate over the relative risks to development associated with a history of maltreatment, lower SES status, and other major risk factors (such as parental psychopathology or environmental stress, see Cicchetti, Taraldson, & Egeland, 1978).

Regarding the general issue of the level of inference appropriate to the present status of theory and research on child maltreatment, it seems that until patterns of emotional expression that reliably distinguish maltreated and nonmaltreated infants can be identified and operationally defined, speculations about the etiology of the patterns and especially about the role of the patterns in initiating or perpetuating maltreatment are premature (Aber & Zigler, 1981).

Nonetheless, the Gaensbauer efforts are very promising due to the theoretical sophistication of the research: their adoption of a transactional model, their focus on component processes of stage-salient development competencies, and their assumption of heterogeneity of cause and of developmental sequelae in child maltreatment (Cicchetti & Rizley, 1981). And in keeping with the old adage that there is nothing so practical as good theory, the most significant long-range contribution of Gaensbauer's work may be its implications for treatment. Analysis of distorted affective communications could become the basis for more effective therapeutic consultations to patients by increasing the clinician's empathy for the parent and by decreasing the parent's powerlessness and frustration in the face of their children's problem behaviors.

2.5. Longitudinal Studies

The only true longitudinal study of the developmental sequelae of child maltreatment has been conducted as part of a prospective study of children at developmental risk due to poverty, mother's age, education, and limited child-care skills and resources. To date, the Minnesota Mother-Child Project has followed 200 children from the third trimester of pregnancy through their second birthday, assessing child characteristics and developmental status, parent and family characteristics, and mother/child interactions at 7 and 10 days, and 3, 6, 9, 12, 18, and 24 months. The outcome measures for child development and mother/child interaction include molar-level and molecular-level assessments across various developmentally salient, age-appropriate tasks: feeding and mother/infant free-play situations at 3 and 6 months (Vaughn, Taraldson, Crichton, & Egeland, 1980), the Bayley Scales of Infant Development at 9 and 24 months, the Strange Situation paradigm for assessing quality of attachment at 12 and 18 months (Ainsworth et al., 1978), and a tool-use/problem-solving paradigm to assess a toddler's emerging autonomy, independent exploration of the environment, and ability to cope with frustration at 24 months (Matas, Arend, & Sroufe, 1978). Observed ratings of mother and infant behaviors and interactions were completed after the 3, 6, and 24 month assessments on such dimensions as baby's responsiveness to mother, disposition, activity level and attention (at 3 and 6 months), and child's frustration toward mother, anger, aggression, noncompliance, positive and negative affect, enthusiasm, coping, and persistence (at 24 months).

From the total sample of 200 children, four groups of maltreated children were identified primarily on the basis of information from home observations of mother/child interaction at 7 and 10 days, and 3, 6, and 12 months coupled with maternal interviews about child-care skills, feelings toward the infant and disciplinary practices, and the completion of a child-care rating scale by the researcher after each home visit. The four groups, defined by maternal behavior and attitude toward their infant, include: physically abusive ($N = 24$), hostile/verbally abusive ($N = 19$), psychological unavailable ($N = 19$), and neglectful ($N = 25$) mothers. According to the authors, classification of the mothers and children based on home observations was further "supported" by systematic observational data from the lab at 9, 12, 18, and 24 months and by welfare, hospital, and public health clinic (the major referral source) case records. Still further validation of the assignment of mother/child pairs to a maltreatment group is cited by the authors: all physically abusive mothers were referred to a child protection agency by someone outside the research project, all neglectful mothers were under the care of either a child protection agency or public health nurse, and all hostile/verbally abusive and psychologically unavailable mothers were classified as such by independent raters who coded mother/child interaction in the lab in limit-setting situations at 12 and 18 months and in the problem-solving situation at 24 months.

Substantial overlap among the groups was reported by Egeland and Sroufe (1981). A majority of verbally abusive (15 of 19), psychologically unavailable (12 of 19), and neglectful (13 of 25) mothers were also classified as physically abusive. Consequently, the major analyses first contrasted the physical-abuse group versus controls, and then the other types of maltreatment with and without physical abuse versus controls. The nonmaltreated children from the sample were used as the control group permitting analyses of the effects of maltreatment independent of class membership.

With one exception, no significant differences were reported between any of the four maltreatment groups and the control group on any of the measures prior to the 12-month assessment. The only exception was that the "psychologically unavailable without physical abuse" group of infants scored higher than the controls on the baby ratings of 3 months, but this difference was washed out by 6 months.

Comparing the physically abusive and control groups, two major differences were identified. At 18 months, the percentage of insecure attachments was 52% for the physical abuse and 29% for the control group. At 24 months, the children from the physical abuse group were rated as more angry, frustrated with mother, noncompliant, and less enthusiastic than controls. In addition, they exhibited a higher frequency of aggressive, frustrated, and noncompliant behaviors and a lower frequency of expressions of positive affect.

The authors state that because the verbally abusive mothers were also classified as physically abusive in 15 of 19 cases, separate analyses of the effects of verbal abuse without physical abuse were not possible. One could also speculate from the data that the physical abuse and verbal abuse of children form a single, unified syndrome or pattern of parent/child interaction during infancy.

The results contrasting the psychologically unavailable groups and the control group are perhaps the most important of the study. Among the infants of psychologically unavailable mothers who were not physically abused, the number of securely attached infants fell from 57% at 12 months to 0% at 18 months, whereas the number of anxious/avoidant attachments rose from 43% to 86%. This compares to a slight rise from 67% to 71% in secure attachments and slight reductions in the number of anxious/avoidant (18% to 16%) and anxious/resistant (15% to 13%) attachments among the control infants from 12 to 18 months. Paralleling this precipitous drop in the percentage of secure attachments is a dramatic decline from 9 to 24 months in Bayley Scale scores for children of psychologically unavailable mothers who were physically abused (112 to 83) and who were not physically abused (118 to 87). Although the control-group children also showed a decline in developmental quotient (DQ) over the same period, the children of psychologically unavailable mothers posted significantly lower DQs at 24 months.

Finally, the percentages of secure attachments among the neglected children with and without physical abuse were 27% and 29%, respectively. Interestingly, the neglected children expressed less positive and more negative affect and posted higher noncompliance, frustration, and anger scores than the controls.

The work of Egeland and Sroufe (1981) clearly illustrates the potential benefits of studying the socio-emotional sequelae of child maltreatment by focusing on age-salient tasks and using a longitudinal design. Whereas other studies failed to identify differences between maltreated children and comparison children from the same social classes, the Minnesota Mother/Child Project findings indicate that maltreated infants and toddlers are less securely attached to their caregivers at 12 and 18 months and exhibit more problem behaviors (e.g., more anger, aggression, frustration, and noncompliance) in a mother-child teaching task at 24 months. By utilizing measures of age-salient developmental tasks, Egeland and Sroufe were also able to differentiate among subtypes of maltreatment by their unique impact on development. The devastating effects of a psychologically unavailable mother on the security of a child's attachment as well as on his or her DQ strongly suggest treating this phenomenon as a unique subtype of maltreatment in infancy and toddlerhood.

The longitudinal results of the study add further evidence to earlier findings of a major divergence among groups of children in developmental paths between the ages of 12 and 24 months. The lack of significant differences between maltreated and comparison infants before 12 months, coupled with the rapid decline in developmental function of the infants of psychologically unavailable mothers between 9 and 24 months, suggest that major shifts in the organization of socio-emotional and behavioral development may accompany the shift from sensorimotor to preoperational intelligence during the second year. In this manner, Egeland and Sroufe are also contributing to the study of major issues in developmental psychopathology as well as to the sequelae of maltreatment: for instance, the issues of continuity/discontinuity of maladaptation and of the impact of pathology in one developmental domain on other domains (Cicchetti & Pogge-Hesse, 1982; Cicchetti & Schneider-Rosen, 1984).

Researchers studying other areas of development, such as social cognition (Selman, 1980), moral development (Kohlberg, 1969), and ego development (Kegan, 1982), have suggested that the careful study of stage transitions is especially important in identifying factors that contribute to group and individual differences in development. The work of Egeland and Sroufe should encourage other researchers in child maltreatment to adopt explicitly longitudinal-developmental strategies and to focus on stage transitions co-occuring across cognitive, behavioral, and socio-emotional domains. For example, whereas the median age of maltreated children reported to state authorities is approximately 6–7 years, researchers studying preschool and early school-age maltreated children have not yet explicitly collected data or interpreted findings in the context of the major shift from preoperational to concrete operational thought. In our culture, this restructing of the organization of thought co-occurs with the emergence of several important socio-emotional developmental tasks: entry into grade school, the need to establish relationships with new adults in positions of authority, increased demands to achieve and to establish peer relationships, and the consolidation of gender identity. The results of the Minnesota Mother/Child Project suggest that research strategies targeting developmental tasks in multiple domains undergoing simultaneous major structural change may be more likely to identify important differences between maltreated and normaltreated children.

Three weaknesses in Egeland and Sroufe's report should also be mentioned. First, their report leaves the reader very confused about how precisely each of the mother/child pairs were assigned to diagnostic groups. Their description raises the possibility that the performance of mother/child pairs on selected outcome measures were used to corroborate group assignments which in turn were used to predict the pair's performance on the same outcome measures. Either the logical flaw needs to be corrected or the authors must achieve greater clarity in their next description. Second, Egeland and Sroufe demonstrate the same tendency as Gaensbauer for studying the important issue of the differential effects of various subtypes of maltreatment by attempting to identify discrete, nonoverlapping subgroups of maltreated children. However, their data indicate substantial overlap of subgroups of children even as it presents evidence that subtypes of maltreatment can be identified. To their credit, the authors did study the effects of subtypes of maltreatment in combination. But a table indicating the total number of maltreated children in the sample and the number of subtypes of maltreatment to which each mother/child pair was assigned would go a long way in correcting the impression that a large number of maltreated children suffer from one and only one major subtype of maltreatment. The Minnesota findings that multiple subtypes of maltreatment often coexist suggest that future researchers also treat the number of subtypes of maltreatment as a major independent/predictor variable. For example, Giovannoni and Becerra (1979) discovered that the single best predictor of the decision by a service agency or court to remove a child from the home was the absolute number of types of maltreatment that a child experienced. Finally, although the authors wished to study the impact of maltreatment independent of the effects of social class and so chose to study a large sample of lower class, high-risk families prospectively, their argument that the subtypes of maltreatment are the *causes* of the developmental differences that they have identified is weakened by their scant treatment of critical family demographic and parental characteristics, which may vary from subtype to subtype. Are psychologically unavailable mothers more often single mothers? Are neglecting mothers from the poorest of poor families? Are physically abusive mothers more prone to other highly erratic behaviors that affect other important areas of family life? If so, the maltreatment alone cannot be said to cause the developmental effects. Developmentalists who are especially sophisticated about "outcome variables" of socio-emotional, behavioral, and cognitive development, their interrelationships and their stage-specific manifestations, should attempt to become as sophisticated in their conceptualization and measurement of independent and/or predictor variables. As Belsky (1980) noted in a review of child-abuse research that attempted to view the phenomenon through the lens of the ecology of human development, patterns of maltreatment may be overdetermined by having causes at multiple levels (societal, familial, individual, and developmental), which interact synergistically. Family, demographic, and parental characteristics are indispensable elements of the comprehensive study of the developmental sequelae of child maltreatment.

3. CONCLUSION

Over the last two decades, researchers of the socio-emotional development of maltreated children have slowly revised earlier research strategies in order to solve problems of the validity and generalizability of earlier findings. The many recent improvements in studies of the socio-emotional development of maltreated children include: specifying the ages, maltreatment subtypes, and demographic characteristics of the children in the studies; deriving questions from developmental theory; and following maltreated children and appropriate comparison children longitudinally.

Despite these improvements, future studies must address a number of remaining issues if they are to form the type of data base needed to guide clinical, legal, and policy decision-making. We will conclude this chapter with a series of recommendations for improving future research on the socio-emotional development of maltreated children. These recommendations are interrelated and are based on our critical review of the last two decades of empirical research and on our own research on the socio-emotional development of maltreated children (Cicchetti & Rizley, 1981).

1. *An overarching, integrative developmental model is needed to guide future studies of the socio-emotional development of maltreated children.*

An examination of the history of the research on the socio-emotional development of maltreated children reveals that little effort has been made to link research on the development of maltreated children with the empirical literature of normal child development. Consequently, the findings of the socio-emotional development of maltreated children are scattered and difficult to integrate into a coherent perspective, primarily because each researcher has tended to focus on an isolated aspect dictated by his or her own theoretical framework.

It is of paramount importance to formulate a developmental model that is broad enough to guide prospective, longitudinal studies of the processes of adaptation in maltreated and nonmaltreated children, and the factors that shape their course. An all encompassing, yet theoretically meaningful model of development, must yield a formulation of developmental continuity that can embrace both change and stability (Reese & Overton, 1970). Additionally, the development model must be sufficiently complex to reflect the multifaceted ways in which constitutional, organismic, and environmental factors affect development. For abused and neglected children in particular, growth and development are related to a plethora of factors, including the type and severity of maltreatment and the quality of care the child receives.

Sameroff and Chandler (1975) have proposed a developmental model that takes into account the interrelationships among dynamic systems and the processes characterizing system breakdown. This "transactional" model views the multiple transactions among environmental forces, caregiver characteristics, and child characteristics as dynamic, reciprocal contributions to the events and outcomes of child development. This model decries the efficacy of simple, linear "cause-effect" models of causality and suggests that it is impossible to understand a child's development by focusing on single pathogenic events. Rather, Sameroff and Chandler (1975) argue that *how* the environment responds to a particular child's characteristics at a particular time must be analyzed in a dynamic fashion. The transactional model presents the environment and the child as mutually influencing. Thus, if a child demonstrates deviant development across time, it is assumed that the child has been involved in a *continuous* maladaptive process. The continued manifestation of maladaptation depends on environmental support, whereas the child's characteristics, reciprocally, determine the nature of the environment.

Sameroff and Chandler's (1975) proposal that we consider a child's position on the "continuum of reproductive casualty" and on the "continuum of caretaking casualty" has obvious applications to the study of the socioemotional development of maltreated children. Since maltreated children are at different points along the continua of "caretaking" and "reproductive" casualty, it is not surprising that we would find both between- and within-group differences in developmental outcome. Thus, the transactional developmental perspective makes it plausible to view maltreatment phenomena as expressions of an underlying dysfunction in the parent-child-environment system, rather than solely the result of aberrant parental personality traits, environmental stressors, or deviant child characteristics (see also, Cicchetti *et al.*, 1978). Since the child and the environment are seen as reciprocally influencing, it follows that behavior at a later point reflects not only the quality of earlier adaptation but also the intervening environmental inputs and supports. As time passes, and as the child develops, the match between child and parent as well as the salient parent characteristics may change. In certain ways the maltreated child creates its own environment and may contribute to its own developmental anomalies. In such a case a declining quality of adaptation would demonstrate continuity of development (Egeland & Sroufe, 1981; Sroufe, 1979a). Such a holistic, biological model is congruent with the theoretical conceptualizations of organismic theorists in the field of developmental psychology (Piaget, 1971; H. Werner, 1948; White, 1976).

We think that this model can help improve both the quality of research on the socio-emotional development of abused and neglected children and the nature of the inferences that are made from the data. In addition, the study of children who are "at risk" for developmental deviations or subsequent pathology can make many significant contributions to our theories of normal development. The investigation of populations where different patterns of socio-emotional development may be expected (as a consequence of enduring and prominent influences that characterize the transaction between the child and the environment), provides a basis for examining claims of universality of a developmental sequence and for affirming and challenging developmental theory while simultaneously contributing precision to current theoretical formulations.

2. *Longitudinal research designs that focus on stage-salient issues must be employed in order to study continuity and discontinuity in the development of maltreated children.*

An accumulating body of empirical work and theoretical formulation has focused on normal processes of development in the affective and social domains and on the relationship between socio-emotional development in childhood and later adaptation (Arend, Gove, & Sroufe, 1979; Main & Weston, 1981; Matas *et al.*, 1978; Sroufe, 1979a,b). Much of this recent research has been guided by the organizational perspective of development (Santostefano & Baker, 1972; Sroufe, 1979b; H. Werner & Kaplan, 1963) and has proceeded in an effort to expand our knowledge of the normal developmental processes of the child. Within the organizational perspective, one studies how behaviors become heirarchically organized into more complex patterns within developmental systems, how later modes and functions evolve from earlier prototypes, and how part functions become integrated into wholes. This organizational approach also refers to relationships among systems—cognitive, social, and emotional—as well as to consequences of advances and lags within one system for other systems (Cicchetti & Serafica, 1981; Cicchetti & Sroufe, 1978; Emde, *et al.*, 1976). It also refers to the consequences of earlier experiences and adaptations for later adaptation (Block & Block, 1980; Erikson, 1959; Sroufe, 1979b).

Of primary concern in the study of socio-emotional development of maltreated children is the search for central organizing principles of social knowledge that are affected by their child-rearing histories and that have implications for their future environmental transactions. The ways in which a child who has been maltreated learns to interact with his or her social milieu has tremendous implications for the development of his or her competence. The issue at stake is the identification of those social factors that may lead to present and future problems in living for the abused or neglected individual. Clearly, if abuse and neglect do have primary impact on socio-emotional development, it becomes important to ask what are the specific, age-appropriate socio-emotional tasks that are most likely to be impaired or disabled as a function of abuse or neglect? The development tasks that are important for a 1 year old are quite different from those of a 5 year old, and we must adjust our assessment strategies accordingly.

Research on normal development from the organizational perspective provides a context for the integration of the disparate findings and suggests focusing on broad-based assessments of stage-salient socio-emotional competencies. Thus, the methodology employed must vary according to the particular age of the child being studied and the nature of the social relationships that are of vital importance at each developmental stage.

The first step in an organizational approach to the study of maltreated children is the determination of age-appropriate issues of and criteria for assessing competence. When development is depicted as a series of behavioral reorganizations around a series of developmental tasks or issues (Greenspan & Lourie, 1981; Sander, 1962; Sroufe, 1979b), it becomes possible to define maladaptation in terms of developmental deviation and to define continuity of adaptation in terms of competence or incompetence across a series of developmentally defined tasks, rather than to pursue static individual consistency. When there are prominent and pervasive disturbances in the transaction between parent and child, the child is at a greater risk for suffering the negative consequences of the "continuum of caretaking casualty" (Sameroff & Chandler, 1975). It is likely that consistent deviations in patterns of parent-child interaction, such as those that characterize the relationship between the maltreated child and the parent, will manifest themselves in the form of difficulties in such salient developmental tasks as achieving a secure attachment relationship with the primary caregiver(s), of expressing affect, of regulating emotional tension, of modulating aggression and anger, of developing social skills, and of maintaining appropriate interpersonal interactions.

Within the organizational framework, continuity refers to the prediction that competence in dealing with one developmental issue (e.g., the formation of a secure attachment relationship) will be related to competence with respect to subsequent issues (e.g., successful integration into and mastery of the peer world). The salient methodological issues are to utilize broadband, age-appropriate situations, each of which elicits a variety of behavioral patterns that are more or less adaptive for that developmental period. We need to use measures that are theoretically meaningful, rather than measures of convenience chosen without a cogent, guiding rationale. Moreover, the measures should be capable of detecting covert or hidden vulnerabilities that may only be apparent under some stress or challenge. Testing the child at the limits is one way of capturing the idea that we need to use progressively more demanding and difficult tasks to detect vulnerabilities or strengths.

Measures that are not demanding enough, that do not test the limits of the child's adaptation, or that tap developmentally inappropriate skills or competencies, may miss the most crucial developmental disabilities.

The recent trend to base research in atypical development upon stage-salient developmental issues identified by research in normal development serves as a reminder that maltreated children are first and foremost children who must face most of the same critical developmental issues and tasks that all other children face. In turn, this focus on stage-specific developmental issues and tasks has encouraged successful studies of continuity between stages in early development in both normal and maltreated children. Contemporary research has begun to construct a picture of how very young maltreated children develop over time. These trends extend the hope that a clearer understanding of those developmental processes effected by maltreatment will encourage the creation of more effective psychological services and prevention efforts for young maltreated children.

The theoretical advances in specifying stage-salient developmental issues have also made it possible to better understand the role of environmental stressors in the etiology of developmental delays and deviations in maltreated children. For instance, the continuity in the quality of mother/child attachment relationships from 12 to 18 months, so stable in middle-class families not experiencing rapid changes in the levels of environmental stress (Waters, 1978), appears to be much less stable in samples of high-risk families who experience variations in levels of stress (Vaughn, Waters, Egeland, & Sroufe, 1979). Refined notions of stage-salient developmental issues are likely to improve our understanding of the ecological context of child maltreatment and to encourage a transactional view of child-and-parent-in-environment as the unit of analysis for developmental studies of child maltreatment.

In light of these recent improvements, we should be optimistic about the challenges that still face scholars of the socio-emotional development of maltreated children. First, most of the improvements just mentioned have been made in research on the development of maltreated infants and toddlers. Before equally productive research can be conducted on the socio-emotional development of older maltreated children, perhaps an integrative theory of stage-salient developmental tasks of the preschool, school-age, and adolescent years is needed to guide research on the effects of maltreatment on later child development. Such a theory will be very difficult to construct and will require solving such puzzles as how to conceptualize and measure the increasing structuralization and organization of personality and behavior and how to conceptualize and measure continuities and discontinuities between earlier and later stages of development. The work of developmental psychologists on ego-control and ego-resiliency (Block & Block, 1980), effectance motivation (Harter, 1978), and children's understanding of interpersonal relationships

(Selman, 1980) provides many valuable clues. But we also require longitudinal research using broadband assessments of stage-salient competencies of older maltreated children in order to develop the kind of empirical picture we are now constructing for younger maltreated children.

Because socio-emotional competence may be the most important and useful predictor of later adaptation and invulnerability to psychopathology (Knight, Roff, Barnett, & Moss, 1979), it is extremely important to initiate large-scale intensive research studies designed to measure and detect both adequacies and disturbances in children who have been maltreated.

3. *Future studies require care and precision in conceptualizing and measuring the range of subtypes of child maltreatment.*

In the past, empirical research efforts on the socio-emotional development of maltreated children attempted to focus on the effects of "pure" subtypes of maltreatment, notably physical abuse. These past efforts recognized the barriers to rigorous research posed by the heterogeneity, potential vagueness, and unreliability of the concept of maltreatment. Using conventional scientific research principles, earlier investigations conceived part of their job to improve empirical studies on the effects of child maltreatment to be focusing on one or two subtypes of maltreatment that were the most important, the most clearly definable, and the most reliably measured.

It now appears that although the decision to operationalize the concept of maltreatment more clearly is laudable, the decision to focus on one or two discrete subtypes of maltreatment may be distorting the true nature of the phenomenon of child maltreatment. The findings of several recent studies challenge the validity of the notion of cases of pure subtypes of maltreatment. Giovannoni and Becerra (1979) developed an 87-item checklist of specific incidents and behaviors indicative of child maltreatment and validated the measure on a sample of 949 representative cases of maltreatment from California. In addition, the seriousness for the child of each incident or behavior was rated by a sample of 159 social workers. From these data, 8 stable, meaningful factors have been derived that describe unique subtypes of maltreatment. The subtypes of maltreatment are normed and listed here in order of their overall seriousness rating (highest to lowest): physical injury, sexual abuse, drug/alcohol abuse (by the child), emotional mistreatment, moral/legal problems (by the adult), failure to provide, child behavior problems, and inadequate physical environment.

Using this system to classify the range of subtypes of maltreatment experienced by a child, Giovannoni and Becerra (1979) found that it was the number of different subtypes of maltreatment, not the most serious single subtype of maltreatment, that was the best predictor of social service agency and court decisions about whether to remove a child from his or her home. In other words, it was the combination of subtypes of maltreatment that most affected clinical and legal decision-making about child placement.

Egeland and Sroufe (1981) presented data that further supported the need to study the full range of subtypes of maltreatment. They found that various combinations of subtypes of maltreatment differentially affect a child's socio-emotional and intellectual development. Thus, the pattern of subtypes of maltreatment that a child experiences not only effects court and service agency decision-making but developmental outcomes as well.

In our own research at the Harvard Child Maltreatment Project (Cicchetti & Rizley, 1981), we have adapted Giovannoni and Becerra's (1979) system of rating subtypes of maltreatment for studies of the developmental sequelae of maltreatment. Based on a preliminary analysis of our data for a subsample of maltreated children between the ages of 4 and 8 years ($N = 32$) who were referred by state social workers to our study, it appears that the children in our sample experienced an average of nearly four unique subtypes of maltreatment (Aber, 1982). A focus on any of these subtypes to the exclusion of others would result in the loss of important information about the nature of risk factors in the child's life. Conceptualizing maltreatment as including a range of unique subtypes that may cooccur, and developing methods that reliably and validly measure this range of subtypes are very important to the quality of future studies of the socio-emotional development of maltreated children.

4. *Future studies should attempt to distinguish clearly the developmental sequelae of maltreatment from the effects of risk factors associated with lower class status and welfare dependency.*

Even a cursory review of the clinical, policy, and empirical literature on child maltreatment indicates a pervasive confusion about the relationship between maltreatment and class. To date, most of the attention has been directed at the issue of the role of lower class status in the etiology of child maltreatment. Some theorists contend that child maltreatment occurs predominantly in families in low socio-economic status (Pelton, 1978) whereas others argue that maltreatment is a phenomenon that knows no class boundaries (Kempe & Kempe, 1978).

But recently theorists and researchers have begun to examine the unique effects of maltreatment on a child's socio-emotional and cognitive development over and above those effects that could be predicted on the basis of the child's lower class status. Elmer's (1977) work, reviewed previously, really brought this issue to a head. On the basis of her work, she concluded that the stress and disorganization of lower class family life exert as negative an influence on later child development as does a history of maltreatment. As discussed earlier, such a conclusion can have a profound impact on social policy. For instance, if maltreatment has no unique effect on the development of children over and above the effects of their lower class status, then perhaps maltreated children do not require special protection under the law or special services to treat their problems and the problems of their families. But such conclusions are valid only if the experimental design of the studies in question logically permits such inferences to be drawn. As we learned in the critical review of Elmer's (1977) work, despite her attempts, issues of class, culture, and related issues of familial access to social and economic resources were intimately intertwined with the issue of how to define child maltreatment.

Today, most empirical studies of the development of maltreated children continue to confound a child's history of maltreatment with class, culture, and familial access to resources. In future studies, greater care should be given to the issue of constructing appropriate comparison groups for empirical studies of the developmental sequelae of child maltreatment. Class, culture, and familial access to resources are all possible rival hypotheses to a history for maltreatment as explanations for group differences in development outcomes as long as studies continue to employ designs that confound these variables.

5. *Future studies should be guided by the notion that there is likely to be heterogeneity in the developmental sequelae of child maltreatment.*

Since there is hetereogeneity in the type(s) of maltreatment that children experience, we assume that there will be heterogeneity in their developmental outcome as well. Children of different ages, at different developmental stages, from diverse environments, and with different experiences, who are exposed to vastly different forms of maltreatment, are likely to manifest competencies, disabilities, and vulnerabilities in a wide variety of specific, age-approprite ways. Accordingly, the assumption of heterogeneity in developmental sequelae has profound implications for the design of future studies.

It is no longer acceptable for researchers to study broad-age ranges of children without conducting developmental analyses of the data. Lumping together children of various developmental levels impedes us from learning about the stage-specific impairments caused by child maltreatment. Moreover, if we are not aware of what children of different developmental stages both can and cannot do, there is a danger in overpathologizing the results. Thus, it is increasingly crucial for researchers to utilize age-appropriate measures and to study stage-salient developmental issues. The combination of an underlying developmental theory to guide research, coupled with increasingly sophisticated measurement and statistical procedures, will enable researchers to resolve the apparent contradictions on the socio-emotional development of maltreated children found in the literature.

Finally, since children most likely will show differential effects of maltreatment, we think that studies should not focus on a single developmental subsystem such as cognitive or emotional development, but rather should examine the interrelationship among various developmental subsystems (see Cicchetti & Schneider-Rosen, 1984). The recent interest in the socio-emotional developmental sequelae of child maltreatment, when coupled with the continued interest in the physical, language, and cognitive developmental sequelae, encourages the view of the maltreated child as a whole child. Similar to Zigler's (1971) reminder to scholars of mental retardation over a decade ago, recent research reminds scholars of child maltreatment that even though the most obvious and easily accessible impairments in a maltreated child's development may be in the physical and cognitive subsystems of development, true understanding of maltreated children as whole children, not simply as sums of their most obvious parts, requires the same careful scrutiny of their socio-emotional, behavioral, and motivational development as well.

6. *Assessments of older children's adaptation to the many new natural environments of their lives are needed in future studies of the socio-emotional developmental sequelae of maltreatment.*

With just a few exceptions, most empirical studies of the socio-emotional development of maltreated children have been conducted either in developmental laboratories or in the children's homes. This is natural enough for the study of young children. For infants and toddlers, the home is *the* major natural environment of their lives and the laboratory can serve as a well-controlled environment for the detailed study of family interactional processes. But as children grow older, the natural contexts of their lives rapidly expand beyond the confines of the home and the family and so can no longer be reasonably well-approximated in the lab. If researchers wish to focus on stage-salient developmental tasks, then future studies of maltreated children need to follow older maltreated children into their enduring environments outside their homes and the laboratory. Theorists and researchers should begin to recognize more fully and to study the increasingly important roles of peer relationships, school life, and community life in the adaptation of maltreated children. Because school is the focus for much of the peer and community life of children from the preschool years through adolescence, school may be an especially important environment in which to study the socio-emotional development of maltreated children.

Future studies would benefit from a research strategy that explicitly attempted to study the same underlying developmental processes and issues of maltreated children in both the laboratory and in natural environments like the school. As Block and Block (1980) have recently argued, researchers can be much more confident of findings confirmed by both experimental test-level and naturalistic observation-level data on the same underlying psychological contruct. The beginnings of such an approach are in place. Studies of maltreated infants and toddlers (George & Main, 1979), and of maltreated preschool and early school-age children (Aber, 1982) point to impairments in maltreated children's relations with novel adults. In infancy and toddlerhood, physically abused children appear to experience greater approach/avoidance conflicts in relation to novel adults (George & Main, 1979). In preschool and early school-age years, maltreated children appear to be especially dependent on the social reinforcement of novel adults (Aber, 1982). These findings raise the question, do such impairments in a child's relations with novel adults affect maltreated children's ability to negotiate entry into nursery school, kindergarten, and elementary school? Future studies of maltreated children that assess the children in the natural environments of their lives can begin to address such questions as whether characteristics of maltreated children identified in laboratory studies, for instance excessive dependency on or physical avoidance of novel adults, may interfere with a child's competent adaptation to the school.

> 7. *Studies of the socio-emotional development of maltreated children must continue to improve in methodological rigor and sophistication and in the logical quality of the inferences drawn from the studies.*

As our review of the last two decades of empirical research indicated, the bulk of studies of the developmental sequelae of child maltreatment continues to contain major weaknesses in research design and method. To cite several of many possible examples, in observational studies of parent/child or child/child interactions, raters are frequently not blind to the children's maltreatment status; in studies attempting to describe the developmental sequelae of maltreatment, no comparison groups or inappropriate comparison groups are employed; even the most carefully designed studies frequently use single, unreliable measures rather than multiple and reliable measures to assess important underlying constructs of development.

Weak designs and flawed methodologies constitute the most serious impediments to improving the quality of research on the socio-emotional development of maltreated children. In turn, errors in design and method repeatedly call into question the validity of the inferences drawn from the work. Yet, clinical and legal decision-making in cases of child maltreatment could benefit from solid developmental studies (Aber, 1980; Aber & Zigler, 1981; Cicchetti & Aber, 1980). Without rigor in design and method, especially in the area of appropriate comparison groups, myth will be put forward in place of knowledge as a guide to social action in the area of child maltreatment (Zigler, 1976).

Even though certain methodological improvements can dramatically enhance the quality of future studies of the developmental sequelae of child maltreatment, no study will ever be methodologically perfect. Studies of phenomena, such as child maltreatment, will always entail unavoidable limits to scientific precision for ethical and practical reasons. There will always be a need to adjust the nature of the inferences drawn from a study to the limits inherent in the study. This critical review identified numerous logical errors in the nature of the inferences drawn from past studies of the developmental sequelae of child maltreatment (for example, see the discussion of Elmer's [1977] study, pp. 169–175 of this review). Due to the increased interest in developmental studies as a source of guidance in clinical, legal, and policy decision-making about child-maltreatment cases, researchers should pay particularly close attention to the nature of the inferences they draw from their own and other researcher's studies and perhaps even attempt to educate the "consumer" of their studies (social workers, pediatricians, lawyers, judges, program planners, and legislators) in the rudimentary logic of their work.

8. *The clinical, legal, and policy implications of studies of the developmental sequelae of maltreatment should be recognized and clarified.*

Despite the fact that a major cause of increased interest in studies of the developmental sequelae of child maltreatment is the clinical, legal, and policy implications of the studies, few researchers have designed their studies to focus on questions with clear, "real-world" implications. We conclude this critical review of research on the socio-emotional development of maltreated children with several examples of how basic developmental research could contribute to the work of clinical, legal, and policy decision-makers.

3.1. Guidelines for Coercive State Intervention

Not all forms of maltreatment justify coercive state intervention into private family life for the purpose of protecting the child. Recently, serious questions have been raised about whether "emotional abuse" or "emotional mistreatment" should be considered a form of maltreatment, which can justify state intervention (Solnit, 1980). Based on a review of the empirical and theoretical literature in psychology as well as certain guiding legal principles, a panel of legal theorists have attempted to describe indices of "emotional damage," which they believe justify coercive state intervention, for example, "extreme withdrawal" or "untoward aggression toward self or others" (Juvenile Justice Standards Project, 1977). Unfortunately, the legal theorists have not adequately recognized that the nature, dynamics, and behavioral expression of responses like withdrawal and aggression vary with the child's developmental level. Consequently, in order for clinical and legal decision-makers to use these indices as guidelines for deciding whether to intervene in cases of "emotional mistreatment" or "emotional harm" requires some familiarity with the stage-specific nature of their expression (Aber & Zigler, 1981). Until the impact of maltreatment on socio-emotional and behavioral development is better understood, clinical, legal, and policy decisionmakers will remain in the dark about what forms of maltreatment cause such severe emotional damage that they serve to justify coercive state intervention.

3.2. Guidelines for the Design and Evaluation of Services to Maltreated Children

With the passage of the Child Abuse Prevention and Treatment Act of 1974, increased interest has developed not only in the treatment of abusive parents but also of abused children. Viewing maltreatment as a transactional phenomenon (Cicchetti & Rizley, 1978, 1981), some clinicians and research-ers began to theorize that the treatment of maltreated children was necessary to help the children avoid continued maltreatment and was also beneficial in remediating the developmental delays caused by maltreatment (Cicchetti *etal.*, 1978; Kempe & Kempe, 1978). But as we described in the first section of this critical review, the clinical descriptions of the psychological deficits and developmental deviations of maltreated children were so contradictory and confusing that they offered little specific advice to clinicians and service planners to design effective child treatment strategies: in short, there was no systematic, empirical data base to help target services to specific and enduring psychological needs of maltreated children. Thus, continued empirical research on the developmental sequelae of maltreatment could greatly aid clinicians and service planners if the research focused on aspects of socio-emotional and behavioral development in a way that also considered how, if deficits or delays in these areas were found, they could become amenable to treatment.

Finally, in times of fiscal austerity, human services in general, including social and psychological services to maltreated children and their families, come under the scrutiny of budget-conscious government administrators and legislators. Increasingly, service providers are asked to document the ben-eficial impact of their service efforts. The inability to provide documentation through evaluation research reports and the like makes the services in ques-tion more vulnerable to the budget-cutter's knife. As we hope this review has made clear, in order to assess the impact of services on the child, one requires stage-specific yardsticks of the child's social, emotional, and behav-ioral development. Thus, basic research on the socio-emotional and behav-ioral development of maltreated children potentially can contribute to the development of the kind of evaluation research methodology required to justify service dollars to skeptical administrators and legislators and to modify ineffective programs to better serve maltreated children and their families. However, the use of basic research by clinical, legal, and policy decision-makers is unlikely without closer collaboration between researchers and decision-makers. Researchers need to become more aware of the types of data that clinicians and policy-makers need to help make decisions and how they would use such data to make decisions. Clinical, legal, and policy decision-makers need to become more aware of the natural limits of any policy-relevant developmental study. Closer collaboration between researchers and decision-makers is in the long-term interests of better research and better service on behalf of maltreated children.

4. REFERENCES

Aber, J. L. Involuntary child placement: Solomon's dilemma revisited. In G. Gerbner, C. Ross, & E. Zigler (Eds.). *Child abuse reconsidered: An analysis and agenda for action.* New York: Oxford University Press, 1980.

Aber, J. L. *The socio-emotional development of maltreated children.* Unpublished doctoral dissertation, Yale University, 1982.

Aber, J. L., & Zigler, E. Developmental considerations in the definition of child maltreatment. *New Directions for Child Development,* 1981, *11,* 1–29.

Ainsworth, M. D. S. The development of mother-infant attachment. In B. M. Caldwell & H. N. Ricciuti (Eds.), *Review of child development research* (Vol. 3). Chicago: University of Chicago Press, 1973.

Ainsworth, M. D. S., Bleher, M., Waters, E., & Wall, S. *Patterns of attachment: A psychological study of the strange situation.* Hillsdale, N. J.: Erlbaum, 1978.

Aragona, J. A., & Eyberg, S. M. Neglected children: Mother's report of child behavior problems and observed verbal behavior. *Child Development,* 1981, *52,* 596–602.

Arend, R., Gove, F., & Sroufe, L. A. Continuity of individual adaptation from infancy to kindergarten: A predictive study of ego resiliency and curiosity in preschoolers. *Child Development,* 1979, *50,* 950–959.

Baharal, R., Waterman, J., & Martin, H. The social-cognitive development of abused children. *Journal of Consulting and Clinical Psychology,* 1981, *49,* 508–516.

Belsky, J. Child maltreatment: An ecological integration. *American Psychologist,* 1980, *35,* 320–335.

Block, J., & Block, J. The role of ego-control and ego-resiliency in the organization of behavior. In W. A. Collins (Ed.), *Minnesota symposia on child psychology* (Vol. 13). Hillsdale, N.J.: Erlbaum, 1980.

Bourne, R., & Newberger, E.H. 'Family autonomy' or 'coercive intervention'? Ambiguity and conflict in the proposed standards for child abuse and neglect. *Boston University Law Review,* 1977, *57,* 670–706.

Burgess, R., & Conger, R. Family interaction in abusive, neglectful, and normal families. *Child Development,* 1978, *49,* 1163–1173.

Caldwell, B. M., Heider, J., & Kaplan, B. *The inventory of home stimulation.* Paper presented at the annual convention of the American Psychological Association, New York, August 1966.

Chance, M. R. S. An interpretation of some agonistic postures: The role of "cut-off" acts and postures. *Symposia of the Zoological Society of London*, 1962, *8*, 71–89.

Cicchetti, D., & Aber, J. L. Abused children—abusive parents: An overstated case? *Harvard Educational Review*, 1980, *50*, 244–255.

Cicchetti, D., & Pogge-Hesse, P. The relation between emotion and cognition in infant development. In M. Lamb & L. Sherrod (Eds.), *Infant social cognition*. Hillsdale, N.J.: Erlbaum, 1981.

Cicchetti, D., & Pogge-Hesse, P. Possible contributions of the study of organically retarded persons to developmental theory. In E. Zigler & D. Balla (Eds.), *Mental retardation: The developmental-difference controversy*. Hillsdale, N.J.: Erlbaum, 1982.

Cicchetti, D., & Rizley, R. *The etiology, transmission, and sequelae of child maltreatment: Toward a three-dimensional threshold transactional model.* Grant proposal, National Center on Child Abuse and Neglect, Administration for Children, Youth and Families, Office of Human Development, 1978.

Cicchetti, D., & Rizley, R. Developmental perspectives on the etiology, intergenerational transmission, and sequelae of child maltreatment. *New Directions for Child Development*, 1981, *11*, 31–55.

Cicchetti, D., & Schneider-Rosen, K. Theoretical and empirical considerations in the investigation of the relationship between affect and cognition in atypical populations of infants. In C. Izard, J. Kagan, & R. Zajonc (Eds.), *Emotions, cognition and behavior*. New York: Cambridge, Press, 1984.

Cicchetti, D., & Serafica, F. The interplay among behavioral systems: Illustration from the study of attachment, affiliation, and wariness in young Down syndrome children. *Developmental Psychology*, 1981, *17*, 36–49.

Cicchetti, D., & Sroufe, L. A. An organizational view of affect: Illustration from the study of Down's syndrome infants. In M. Lewis & L. Rosenblum (Eds.), *The development of affect*. New York: Plenum Press, 1978.

Cicchetti, D., Taraldson, B., & Egeland, B. Perspectives in the treatment and understanding of child abuse. In A. Goldstein (Ed.), *Prescriptions for child mental health and education*. New York: Pergamon Press, 1978.

Cohn, A. An evaluation of three demonstration child abuse and neglect treatment programs. *Journal of the American Academy of Child Psychiatry*, 1979, *18*, 283–291.

Costanzo, P. E., Coie, J. D., Grumet, J. F., & Farnill, D. A reexamination of the effects of intent and consequence on children's moral judgments. *Child Development*, 1973, *44*, 154–161.

Disbrow, M. A., Doerr, H, & Caulfield, C. Measuring the components of parents' potential for child abuse and neglect. *Child Abuse and Neglect*, 1977, *1*, 279–296.

Egeland, B. Preliminary results of a prospective study of the antecedents of child abuse. *Child Abuse and Neglect*, 1979, *3*, 269–278.

Egeland, B., & Sroufe, L. A. Developmental sequelae of maltreatment in infancy. *New Directions for Child Development*, 1981, *11*, 77–92.

Elmer, E. *Children in jeopardy: A study of abused minors and their families.* Pittsburgh: University of Pittsburgh Press, 1967.

Elmer, E., *Fragile families, troubled children.* Pittsburgh: University of Pittsburgh Press, 1977.

Elmer, E., & Gregg, G. S. Developmental characteristics of abused children. *Journal of Pediatrics*, 1967, *40*, 596–602.

Emde, R. N., Gaensbauer, J., & Harmon, R. *Emotional expression in infancy: A biobehavioral study.* New York: International Universities Press, 1976.

Emde, R. N., Kligman, D. H., Reich, J. G., & Wade, T. D. Emotional expression in infancy: Initial studies of social signaling and an emergent world. In M. Lewis & L. Rosenblum (Eds.), *The development of affect*. New York: Plenum Press, 1978.

Emde, R. N., Katz, E. L., & Thorpe, J. K. Emotional expression in infancy: Part II. Early deviations in Down's syndrome. In M. Lewis & L. Rosenblum (Eds.), *The development of affect*. New York: Plenum Press, 1978.

Erikson, D. H. Identity and the lifecycle. *Psychological issues* (Vol. 1, No. 1). New York: International Universities Press, 1959.

Erikson, E. H. *Childhood and society* (2nd ed.). New York: Norton, 1950.

Eyberg, S., Robinson, E., Kniskern, J., & O'Brien, P. *Manual for coding dyadic parent-child interactions*. Unpublished manuscript, University of Oregon Health Sciences Center, 1978.

Feshbach, N. D. The effects of violence in childhood. *Journal of Clinical Child Psychology*, 1973, *2*, 28–31.

Feshbach, S. Aggression. In P. H. Mussen (Ed.), *Carmichael's manual of child psychology* (Vol. 2). New York: Wiley, 1970.

Flavell, J. H., Botkin, P. T., Fry, C. I., Wright, J. W., & Jarvis, P. E. *The development of role-taking and communication skills in children*. New York: Wiley, 1968.

Fontana, V. J. *Somewhere a child is crying: Maltreatment—causes and prevention*. New York: Macmillan, 1973.

Friedman, S., & Morse, C. Child abuse: A five-year follow-up of early case finding in the emergency department. *Pediatrics*, 1974, *54*, 404–410.

Friedrich, W. N., & Boriskin, J. A. The role of the child in abuse: A review of the literature. *American Journal of Orthopsychiatry*, 1976, *46*, 580–590.

Gaensbauer, T. J., & Sands, S. K. Distorted affective communications in abused/neglected infants and their potential impact on caretakers. *Journal of the American Academy of Child Psychiatry*, 1979, *18*, 236–250.

Gaensbauer, T. J., Mrazek, D., & Harmon, R. J. Affective behavior patterns in abused and/or neglected infants. In N. Frude (Ed.), *The understanding and prevention of child abuse: Psychological approaches*. London: Concord Press, 1980.

Gaines, R., Sangrund, A., Green, A. H., & Power, E. Etiological factors in child maltreatment: A multivariate study of abusing, neglecting, and normal mothers. *Journal of Abnormal Psychology*, 1978, *87*, 531–540.

Galdston, R. Observations of children who have been physically abused by their parents. *American Journal of Psychiatry*, 1965, *122*, 440–443.

Galdston, R. Violence begins at home: The parents' center project for the study and prevention of child abuse. *American Academy of Child Psychiatry*, 1971, *10*, 336–350.

Galdston, R. Preventing the abuse of little children: The parents' center project for the study and prevention of child abuse. *American Journal of Orthopsychiatry*, 1975, *45*, 372–381.

Garmezy, N. Children at risk: The search for the antecedents of schizophrenia. Part I. Conceptual models and research methods. *Schizophrenia Bulletin*, 1974, *8*, 13–89.

George, C., & Main, M. Social interactions of young abused children: Approach, avoidance, and aggression. *Child Development*, 1979, *50*, 306–318.

George, C., & Main, M. Abused children: Their rejection of peers and caregivers. In T. M. Field, S. Goldberg, D. Stern, & A. M. Sostek (Eds.), *High risk infants and children: Adult and peer interactions*. New York: Academic Press, 1980.

Giovannoni, J. M., & Becerra, R. M. *Defining child abuse*. New York: Free Press, 1979.

Goldstein, J., Freud, A., & Solnit, A. *Beyond the best interests of the child*. New York: Free Press, 1973.

Goldstein, J., Freud, A., & Solnit, A. *Before the best interests of the child*. New York: Free Press, 1979.

Gordon, A. *Patterns and determinants of attachment in infants with the non-organic failure to thrive syndrome*. Unpublished doctoral dissertation. Harvard Graduate School of Education, 1979.

Gray, J., & Kempe, R. S. The abused child at time of injury. In H. P. Martin (Ed.), *The abused child: A multi-disciplinary approach to developmental issues and treatment*. Cambridge, Mass.: Ballinger, 1976.

Green, A. H. Self-destructive behavior in physically abused schizophrenic children. *Archives of General Psychiatry*, 1968, *19*, 171–179.

Green, A. H. Psychopathology of abused children. *Journal of the American Academy of Child Psychiatry*, 1978, *17*, 92–103. (a)

Green, A. H. Self-destructive behavior in battered children. *American Journal of Psychiatry*, 1978, *135*, 579–583. (b)

Greenspan, S., & Lourie, R. Developmental structuralist approach to the classification of adaptive and pathologic personality organizations: Infancy and early childhood. *American Journal of Psychiatry*, 1981, *138*, 725–735.

Gregg, G. S., & Elmer, E. Infant injuries: Accident or abuse? *Pediatrics*, 1969, *44*, 434–439.

Harter, S. Effectance motivation reconsidered: Toward a developmental model. *Human Development*, 1978, *21*, 34–64.

Herrenkohl, E. C., & Herrenkohl, R. C. A comparison of abused children and their nonabused siblings. *Journal of the American Academy of Child Psychiatry*, 1979, *18*, 260–269.

Herrenkohl, R. C., & Herrenkohl, E. C. Some antecedents and developmental consequences of child maltreatment. *New Directions for Child Development*, 1981, *11*, 57–76.

Hoffman, M. Power assertion by the parent and its impact on the child. *Child Development*, 1960, *31*, 129–143.

Hollingshead, A. B. *Two-factor index of social position*. Unpublished manuscript, Yale University, 1956.

Holter, J. C., & Friedman, S. B. Child abuse: Early case finding in the emergency department. *Journal of Pediatrics*, 1968, *42*, 128–138. (a)

Holter, J. C. & Friedman, S. B. Principles of management in child abuse cases. *American Journal of Orthopsychiatry*, 1968, *38*, 127–136. (b)

Johnson, B., & Morse, H. A. Injured children and their parents. *Children*, 1968, *15*, 147–152.

Juvenile Justice Standards Project. *Standards relating to abuse and neglect*. Cambridge, Mass.: Ballinger, 1977.

Kegan, R. *The evolving self*. Cambridge: Harvard University Press, 1982.

Kempe, R., & Kempe, C. H. *Child abuse*. London: Fontana/Open Books, 1978.

Kent, J. A follow-up study of abused children. *Journal of Pediatric Psychology*, 1976, *1*, 25–31.

Kinard, E. M. Emotional development in physically abused children. *American Journal of Orthopsychiatry*, 1980, *50*, 686–696.

Knight, R., Roff, J., Barnett, J., & Moss, J. Concurrent and predictive validity of thought disorder and affectivity: A 22-year follow-up. *Journal of Abnormal Psychology*, 1979, *88*, 1–12.

Kohlberg, L. *Stages in the development of moral thought and action*. New York: Holt, Rinehart & Winston, 1969.

Main, M. Analysis of a peculiar form of reunion behavior seen in some daycare children: Its history and sequelae in children who are home-reared. In R. Webb (Ed.), *Social development in childhood: Daycare programs and research*. Baltimore: Johns Hopkins University Press, 1977.

Main, M. Avoidance in the service of proximity. In K. Immelmann, G. Barlow, M. Main, & L. Petrinovich (Eds.), *Behavioral development: The Bielefeld interdisciplinary project*. New York: Cambridge University Press, 1980.

Main, M., & Weston, D. Security of attachment to mother and father: Related to conflict behavior and the readiness to establish new relationships. *Child Development*, 1981, *52*, 932–940.

Martin, H. P. (Ed.). *The abused child: Multidisciplinary approach to developmental issues and treatment*. Cambridge, Mass.: Ballinger, 1976.

Martin, H. P., & Beezeley, P. Personality of abused children. In H. P. Martin (Ed.), *The abused child*. Cambridge, Mass.: Ballinger, 1976.

Martin, H. P., & Beezeley, P. Behavioral observations of abused children. *Developmental Medicine and Child Neurology*, 1977, *19*, 373–387.

Martin, H. P., Beezeley, P., Conway, E. F., & Kempe, C. H. The development of abused children. *Advances in Pediatrics*, 1974, *21*, 25–73.

Matas, L., Arend, R. A., & Sroufe, L. A. Continuity of adaptation in the second year: The relationship between the quality of attachment and later competence. *Child Development*, 1978, *49*, 547–556.

Mischel, W. *Personality and assessment*. New York: Wiley, 1968.

Mischel, W., Zeiss, R., & Zeiss, A. Internal-external control and persistence: Validation and implications of the Stanford Preschool Internal-External Scale. *Journal of Personality and Social Psychology*, 1974, *29*, 265–278.

Morse, C. W., Sahler, O., & Friedman, S. A three-year follow-up study of abused and neglected children. *American Journal of Diseases of Children*, 1970, *120*, 439–446.

Mueller, C. W., & Parcel, T. L. Measures of socio-economic status: Alternatives and recommendations. *Child Development*, 1981, *52*, 13–30.

Newberger, C. M. *Parental conceptions and child abuse: A developmental approach to clinical understanding*. Paper presented at the meeting of the Society for Research in Child Development, San Francisco, March 1979.

Newberger, E. H., Reed, R., Daniel, J., Hyde, J., & Kotelchuck, M. Pediatric social illness: Toward an etiological classification. *Pediatrics*, 1977, *60*(2), 178–185.

Ounsted, C., Oppenheimer, R., & Lindsay, J. Aspects of bonding failure: The psychopathological and psychotherapeutic treatment of families of battered children. *Developmental Medicine and Child Neurology*, 1974, *16*, 447–456.

Patton, R., & Gardner, C. Influence of family environment on growth. *Pediatrics*, 1962, *30*, 957–962.

Pavenstadt, E. A comparison of the child-rearing environment of upper-lower and very lower-lower class families. *American Journal of Orthopsychiatry*, 1965, *35*, 84.

Pelton, L. Child abuse and neglect: The myth of classlessness. *American Journal of Orthopsychiatry*, 1978, *48*, 608–617.

Piaget, J. *Biology and knowledge*. Chicago: University of Chicago Press, 1971.

Piaget, J., & Inhelder, B. *The psychology of the child*. New York: Basic Books, 1969.

Prugh, D., & Harlow, R. Masked deprivation in infants and young children. In *Deprivation of maternal care* (WHO Public Health Paper No. 14). Geneva: World Health Organization, 1962.

Quay, H. C., & Peterson, D. R. *Manual for the behavior problem checklist*. Champaign: University of Illinois, Children's Research Center, 1967.

Reese, H. & Overton, W. Models of development and theories of development. In L. R. Goulet & P. Baltes (Eds.), *Life-span developmental psychology: Research and theory*. New York: Academic Press, 1970.

Reid, J. B., & Taplin, P. S. *A social interactional approach to the treatment of abusive families*. Unpublished manuscript, Oregon Research Institute, Eugene, Oregon, 1977.

Reidy, T. J. The aggressive characteristics of abused and neglected children. *Journal of Clinical Psychology*, 1977, *33*, 1140–1145.

Relich, R., Giblin, P., Starr, R., & Agronow, S. Motor and social behavior in abused and control children: Observations of parent-child interactions. *The Journal of Psychology*, 1980, *106*, 193–204.

Rolston, R. The effect of prior physical abuse on the expression of overt and fantasy aggressive behavior in children. *Dissertations Abstracts International*, 1971, *32*, 3016A.

Rothenberg, B. Children's social sensitivity and the relationship to interpersonal competence, intrapersonal comfort, and intellectual level. *Developmental Psychology*, 1970, *2*, 335–350.

Rutter, M. Early sources of security and competence. In J. S. Bruner & A. Garton (Eds.), *Human growth and development*. London: Oxford University Press, 1978.

Sameroff, A. J. Early influences on development: Fact or fancy? *Merill-Palmer Quarterly*, 1975, *21*, 267–294.

Sameroff, A., & Chandler, M. Reproductive risk and the continuum of caretaking casualty. In F. Horowitz (Ed.), *Review of child development research* (Vol. 4). Chicago: University of Chicago Press, 1975.

Sander, L. Issues in early mother-child interaction. *Journal of the American Academy of Child Psychiatry*, 1962, *1*, 141–166.

Santostefano, S., & Baker, A. H. The contribution of developmental psychology. In B. Wolman (Ed.), *Manual of child psychopathology*. New York: Wiley, 1972.

Sears, R. R., Maccoby, E. E., & Levin, H. *Patterns of child-rearing*. Evanston, Ill.: Row, Peterson, 1957.

Selman, R. L. *The growth of interpersonal understanding: Developmental and clinical and analyses*. New York: Academic Press, 1980.

Silver, H., & Finkelstein, M. Deprivation dwarfism. *Journal of Pediatrics*, 1967, *70*, 317–324.

Slosson, R. L. *Slosson intelligence test for children and adults*. East Aurora, New York: Slosson Educational Publications, 1961.

Solnit, A. Too much reporting, too little service: Roots and prevention of child abuse. In G. Gerbner, C. Ross, & E. Zigler (Eds.), *Child abuse: An agenda for action*. New York: Oxford University Press, 1980.

Spitz, R. *The first year of life*. New York: International Universities Press, 1965.

Sroufe, L. A. The coherence of individual development: Early care, attachment and subsequent developmental issues. *American Psychologist*, 1979, *34*, 834–841. (a)

Sroufe, L. A. Socioemotional development. In J. Osofsky (Ed.), *Handbook of infant development*. New York: Wiley, 1979. (b)

Stern, D. *The first relationship*. Cambridge: Harvard University Press, 1977.

Terr, L. C. A family study of child abuse. *American Journal of Psychiatry*, 1970, *127*, 125–131.

Vaughn, B., Waters, E., Egeland, B., & Sroufe, L. A. Individual differences in infant-mother attachment at twelve and eighteen months: Stability and change in families under stress. *Child Development*, 1979, *50*, 971–975.

Vaughn, B., Taraldson, B., Crichton, L., & Egeland, B. Relationships between neonatal behavioral organization and infant behavior during the first year of life. *Infant Behavior and Development*, 1980, *3*, 47–66.

Wald, M. State intervention on behalf of "neglected" children: A search for realistic standards. *Stanford Law Review*, 1975, *27*, 985–1040.

Wald, M. State intervention on behalf of "neglected" children: Standards for removal of children from their homes, monitoring the status of children in foster care, and termination of parental rights. *Stanford Law Review*, 1976, *28*, 623–707.

Waters, E. The stability of individual differences in infant-mother attachment. *Child Development*, 1978, *49*, 483–494.

Watson, M. *A developmental sequence of role-playing*. Paper presented at the meeting of the American Psychological Association, San Francisco, August 1977.

Werner, E., & Smith, R. *Vulnerable but invincible: A study of ego-resilient children*. New York: McGraw-Hill, 1982.

Werner, H. *Comparative psychology of mental development*. Chicago: Follett, 1948.

Werner, H., & Kaplan, B. *Symbol formation*. New York: Wiley, 1963.

White, S. H. The active organism in theoretical behaviorism. *Human Development,* 1976, *19,* 99–107.

Zigler, E. The retarded child as a whole person. In H. E. Adams & W. K. Boardman III (Eds.), *Advances in experimental clinical psychology.* New York: Pergamon Press, 1971.

Zigler, E. Controlling child abuse in America: An effort doomed to failure? In R. Bourne & E. H. Newberger (Eds.), *Critical perspectives on child abuse.* Lexington, Mass.: Lexington Books, 1979.

Developing Identity
Ego Growth and Change during Adolescence

STUART T. HAUSER AND DONNA J. FOLLANSBEE

1. INTRODUCTION

1.1 Definitions of Ego Identity and Ego Development

> What the regressing and growing, rebelling and maturing youths are
> now primarily concerned with is who and what they are . . . and
> how to connect the dreams, idiosyncrasies, roles and skills cultivated
> earlier with the occupational and sexual prototypes of the day.
>
> —Erikson, 1963, p. 307

Building on such theorists as Hartmann (1958) and H. S. Sullivan (1953), Erikson has contributed greatly to our current understanding of adolescent development. Early interest in the adolescent ego centered on depicting ways that ego defenses warded off the tumultuous upsurge of pubertal impulses (A. Freud, 1936, 1958; Josselyn, 1952). As noted by Anna Freud:

> The physiological process which marks the attainment of physical sexual maturity
> is accompanied by a stimulation of instinctual processes. . . . Aggressive impulses
> are intensified to the point of complete unruliness, hunger becomes voracity and
> the naughtiness of the latency period turns into criminal behavior of adolescence.
> (1936, p. 40)

In more recent formulations turmoil and upsurge are taken into account, but the emphasis has shifted toward autonomous ego functions, mastery, adaptation, and identity formation.[1]

[1]Consistent with this shift are the findings reported by Offer. A major theme in the work of Offer and his associates (Offer, 1969; Offer & Offer, 1975) is that "turmoil" does not appear as the predominant mode in their studies of (normal) adolescents.

Stuart T. Hauser and Donna J. Follansbee • Department of Psychiatry, Harvard Medical School, Boston, Massachusetts 02115.

The adolescent's task is not seen by Erikson as restricted to developing multiple drive-mediating functions; rather, the task is broadly defined in terms of developing an *ego identity* (Bourne, 1978a,b; Erikson, 1963, 1956). This view is an epigenetic one, in which the struggle to achieve ego identity is one stage in the total process of personality development. As Erikson (1959) states, the formation of ego identity during adolescence "bridges the early childhood stages, when body and parent images were given their specific meaning, and the later stages, when a variety of social roles become available and increasingly coercive" (p. 50).

Josselson (1980), in her recent review, summarizes Erikson's contributions to the understanding of adolescence:

1. *Ego continuity.* Adolescence is intimately linked to the rest of the life cycle, with its own special tasks that have been in preparation all along.
2. *Adolescence as a maturational necessity.* Adolescence is a psychological demand that will be imposed in the individual whether or not there is internal push for it.
3. *Ego identity.* Ego integration at adolescence is an emergent phenomenon in the sense that the organization of aspects of self is more than the sum of parts and resides precisely in the manner in which the parts are synthesized. (p. 189, italics added)

By placing the period of adolescence *and* the task of ego identity formation within a developmental sequence of stages, Erikson has provided a foundation for other theorists. Dual goals of this chapter are to describe several of these perspectives and to then apply them to observations and questions about psychosocial development in adolescence. Before reviewing these frameworks in detail, we introduce the reader to them through several key concepts.

The work of Marcia (1966, 1967, 1976a,b) began as an attempt to operationalize Erikson's notion of ego identity and has become a widely used approach in empirical studies of "identity status" (Marcia, 1980). Marcia studies the *conscious* properties of the adolescent identity crises; for Erikson, the resolution of this complex crisis is largely unconscious (White & Speisman, 1977). Marcia focuses on decision-making and personal commitment in the areas of occupation and ideology. He discusses four modes of identity development ("identity status") characteristic of late adolescence: identity achievement, identity foreclosure, identity diffusion, and moratorium (Marcia, 1967, 1976a,b). These modes illustrate Marcia's view of ego identity as an underlying process that leads to the patterning of a self-constructed, dynamic organization of abilities, drives, needs, and beliefs (Marcia, 1980). The identity statuses depict the various patterns that are outcomes of the process. Marcia defines the patterns through focusing on conscious thoughts and attitudes toward work and ideology.

In contrast to Marcia's emphasis on conscious ideas, Hauser (1971, 1976b) stresses the study of identity in terms of underlying self-image processes. These studies assess self-image patterns along lines of integration and continuity. In reviewing the results of this work and the identity–self-image studies of others, we will also consider broader questions concerning self-concepts and their relationship to adolescent development.

Block (1971, 1981) also highlights the topic of continuity in development in his studies of the psychological growth of individuals. He studied individuals who, as children, had taken part in an investigation by Mac-Farlane (1938). He evaluated many of the subjects in their thirties and again in their forties. Through analysis of the subjects as adolescents and adults, Block constructs a personality typology that covers behavior, adjustment, and individual differences for each specific type (J. Block, 1981).

Central to Block's perspective of personality type are the dimensions of "ego resiliency" and "ego control" as organizing developmental constructs (J. Block & Block, 1979). Block's notions are similar to Erikson's: they emphasize the continuity of the ego in development. However, an important difference is that Block discusses the antecedents and consequences of ego development in terms of a *personality typology*, rather than the ego stages described by Erikson.

We will also review the theoretical and psychometric contributions of Loevinger. Her work, though similar to Erikson's in its emphasis on processes and stages, provides still another perspective. In Loevinger's view, the concept of ego refers to the individual's integrative processes and overall frame of reference (Hauser, 1976a). Her model elaborates a developmental continuum of variously structured ego stages. Each stage represents new ways of organizing experience: "a more differentiated perception of self, of the social world, and of the relations of one's feelings and thoughts to those of others" (Candee, 1974, p. 621). Loevinger's perspective is a particularly important and unique one. In addition to identifying stages of ego development, she has developed a theoretically linked assessment technique. Although similar to Erikson's in describing developmental stages, Loevinger's model is also based on empirical observations and is most amenable to systematic research.

Thus far, we have presented a general orientation and overview of the theories to be discussed. We have begun to sketch the meaning of the terms *ego development* and *ego identity* as contained within the respective theories. In the following sections, we look more closely at these theories with respect to *adolescent development*. We examine such topics as stages of ego development and styles of identity crisis resolution in normal development. We also attend to broader subjects, such as the relationship of ego development to self-concept, and the familial contexts that foster various types of ego growth.

2. SELECTED THEORIES AND THEIR MEASURES

2.1. Ego Identity

2.1.1. Erikson's Theory

In Erikson's formulations, the more traditional psychoanalytic theory is expanded to incorporate social dimension and to extend into adult life. In essence, Erikson posits that personality development follows an epigenetic sequence. Thus, as the individual matures, he or she passes through various stages at the center of which are specific psychosocial crises. The resolution of each crisis enables the individual to progress to the next stage and its corresponding crisis. Crisis resolution requires an interaction of the individual with significant people and societal institutions. The result is both formation of identity and integration of the individual within the society.

Within Erikson's framework are several important constructs. The most central is the *crisis*. The term is used

> in a developmental sense to connote not a threat or catastrophe but a turning point, a crucial period of increased vulnerability and heightened potential, and therefore the ontogenetic source of generational strength and maladjustment. (Erikson, 1963, p. 294)

Erikson's definition of crisis is reminiscent of the ancient conception of crisis as both a danger and a challenge: each successive crisis contains potential for maladjustment as well as for growth.

In the adolescent stage of "identity versus role diffusion," the concept of crisis is especially evident. The danger is to remain in role diffusion, without a clear sense of self or place in the world. Successful resolution of this stage's challenge results in the formation of *ego identity*, a personality configuration that is

> more than the sum of childhood identifications. It is the accrued experience of the ego's ability to integrate these identifications with the vicissitudes of the libido, with the aptitudes developed out of endowment, and with the opportunities afforded in social roles. The sense of ego identity, then, is the accrued confidence that the inner sameness and continuity are matched by the sameness and continuity of one's meaning for others. (Erikson, 1950, p. 228)

Ego identity, then, refers to a subsystem of the ego that integrates self-representations derived from childhood, and that facilitates resolution of the psychosocial adolescent identity crisis (Erikson, 1959).

The *identity crisis* occurs during adolescence. This is the time in which society itensifies in the adolescent a concern with self-acceptance and, hopefully, offers the recognition that enables the youth to develop an optimal sense of *identity*. This identity is "a feeling of being at home in one's body, a sense of knowing where one is going and an inner assuredness of anticipated recognition from those who count" (Erikson, 1968, p. 165). The opposite of identity is *role diffusion,* that is, an inability to feel confident in one's sense of self, occupational purpose, or sexuality. The process of identity formation requires a period of *moratorium*—a time in which the adolescent is free to try on various roles, to remain uncommitted to any particular ideology, and to develop his or her own sexuality. The time of moratorium as well as the "identity crisis" itself is an important point during lifelong identity formation; it is neither the beginning nor the end. The sense of identity that evolves prepares the adolescent for the concerns of the next stage: further occupational commitment and interpersonal intimacy. The development that occurs as the adolescent gains a sense of identity is an active focus of research, well-exemplified in the work of Marcia.

2.1.2. Marcia's Paradigm: Identity Status

Recently, Marcia operationalized the constructs of "ego identity" and the "identity crisis." He defines four modes of "identity statuses." Each mode represents the degree of crisis and commitment present in the adolescent's formation of occupational goals and ideological beliefs. *Crisis* refers to the adolescent's engagement in a decision-making period. *Commitment* is the extent of personal investment the adolescent exhibits. The interaction of the two characterizes the type of resolution the adolescent achieves.

The four types of identity status are: identity achievement, foreclosure, diffusion, and moratorium. Identity achievement and identity diffusion are the polar forms of identity resolution in Erikson's theory. In Marcia's terms, *identity achievers* are those who have experienced a decision-making period, have developed occupational choices, and have made decisions on their own. The ideological beliefs of these adolescents have also been evaluated and reaffirmed or discarded in support of their current views and behavior. Those in the *identity diffusion* status, however, lack all form of commitment. Although they may have experienced a crisis period, they have not yet made occupational choices and are not concerned about them. Their ideological beliefs are either totally unformed, or are a potpourri of various thoughts and opinions, with no underlying philosophy. *Moratorium* adolescents resemble those of diffused identity in their lack of commitment and vague ideology. However, they are concerned about their lack of direction and are actually in the midst of the identity crisis, trying to compromise among parental wishes, society's demands, and their own capabilities. At times they may appear bewildered, due to their internal preoccupation with what seem to be difficult, weighty questions (Marcia, 1980, p. 161). *Foreclosure* adolescents do not reveal uncertainty. They express very definite commitments, yet they have *not* experienced crisis of any sort. Their goals are identical to those upheld by their parents, and most experiences serve to reaffirm childhood beliefs.

2.1.3. Measurement and Application

The method used to determine an identity status is a semistructured interview, with questions in three areas: occupation, religion, and politics.[2] In each of three sections, probes are designed to ascertain the adolescent's previous or present experience of "crisis" and ensuing commitments to occupational and to religious and political ideology (Bourne, 1978a). Each area is rated separately and then aggregate identity status is assigned. Examples of characteristic responses for the four statuses can be seen in the answers to the interview question, "How willing do you think you'd be to give up _____, if something better came along?" Following each status "type' we give a typical reply:

1. *Identity achievement:* "Well, I might, but I doubt it. I can't see what something better would be for me."
2. *Moratorium:* "I guess if I knew for sure I could answer that better. It would have to be something in the general area—something related."
3. *Foreclosure:* "Not very willing. It's what I've always wanted to do. The folks are happy with it and so am I."
4. *Identity diffusion:* "Oh, sure, if something better came along, I'd change just like that" (Marcia, Scoring Manual for Ego Identity Status, 1964).

Numerous studies applying the identity status interview have appeared in the past 17 years. Strong evidence for the technique's reliability and validity has been reported (Adams & Shea, 1979; Marcia, 1976a; Schenkel & Marcia, 1972; A. S. Waterman & Waterman, 1972). Reviews by Bourne (1978a,b) and Marcia (1980) cover this interesting work. Since two sets of studies have clinical relevance, we summarize their findings here.

2.1.3a. Interpersonal Relations. Some of the research focuses on the interaction styles of various identity status adolescents and their "level" of interpersonal intimacy. Marcia (1976a) suggests that the statuses may be thought of as being in a temporal sequence, progressing from diffusion to achievement. Adolescents nearest the status of identity achievement have been found to have the greatest capacity for intimacy. Those still in identity diffusion have more distant, less personal relationships (Constantinople, 1969; Kinsler, 1972).

---

[2]In a recent paper, Grotevant, Thorbecker, & Meyer (1982) describe an expansion of this interview (and coding system) to include the interpersonal domain (friendships, dating, and sex roles).

Orlofsky, Marcia, and Lesser (1973), using this "intimacy status," reports empirical indications of a strong link between identity achievement and the capacity for intimacy. Subjects further along in the process of identity acquisition were found to have the most intimate relationships (Orlofsky *et al.*, 1973). Those in the foreclosure and diffusion statuses were more likely to have stereotyped, superficial relationships. Nearly one third of the identity-diffusion adolescents were characterized by fewer relationships in general and a tendency toward self-preoccupation (Orlofsky *et al.*, 1973).

J. M. Donovan (1975) gives a more clinical description, based on intensive interviews and on psychological tests. He found identity-diffusion adolescents to be isolated, withdrawn, and wary of both peers and authorities. Those classified as foreclosure were loving and affectionate, but not apt to show any strong feelings—positive or negative. Moratorium subjects thrived on intense relationships. Their interactions with others were characterized by ambivalence and competitiveness. The identity achievers were seen as having a capacity for caring for others in a relaxed, nonbinding way. These findings suggest that an adolescent's progression toward identity acquisition is reflected by quality of interpersonal relations. Moreover, they provide support for Erikson's theory, where the capacity for intimacy underlines the importance of considering ego identity and the ability to relate to others as part of an integrated developmental sequence.

2.1.3b. Identity in Females. Much of Marcia's original research was carried out only with males. Subsequent work revealed different patterns for women and led to a revision of Marcia's original interview, the addition of focusing on "attitudes toward premarital intercourse" (Schenkel & Marcia, 1972). The revision was based on Erikson's (1968) suggestion that some combination of social relationships, marriage, and sexual attitudes are important issues in female identity formation. One finding derived from the new interview is that more females than males have gone through crises and made commitments in the area of sexuality (Marcia, 1980).

Erikson views the female's emphasis on social and sexual issues as preparation for eventual wife/mother roles. Other authors suggest that this emphasis reflects women's greater concern with interpersonal relatedness. Hence, for women, their own intrinsic emphasis may call for a different type of identity crisis. We return to this theme in a subsequent section, when we consider implications for clinical application and further research.

Another relevant finding points to sex differences in the linking of presumably separate identity statuses. Men who are in the identity achievement and moratorium status perform similarly in a number of areas, including intimacy in relationships, manipulability of self-esteem, locus of control, and concept attainment under stress (Marcia, 1980; C. K. Waterman, Bruebel, & Waterman, 1970). However, for women, those in the identity achievement and foreclosure statuses are more alike. For example, identity achievement and foreclosure women choose more difficult college majors (Marcia & Friedman, 1970); are less conforming and more comfortable in social situations (Toder & Marcia, 1973); and have more positive "adjustment" scores (Marcia, 1980) than moratorium and identity-diffusion women.

One way to understand these sex differences is along the lines of adaptation. Marcia (1976b) suggests that *stability* (which characterizes both foreclosure and identity achievement) is more significant for women, whereas *temporal proximity* (which groups moratorium and identity achievement) is more important for men. He argues that men receive much societal support for enduring the unstable period of crisis and commitment, leading them to identity achievement and preparing them for an occupational and ideological place in society. On the other hand, women may be encouraged to bypass the confusion of the identity crisis and remain conventionally foreclosed (Marcia, 1976b).

Recent work emphasizes the complexity of sex differences in identity development and leads to a second, related theme. Orlofsky (1978) reports that the men and women rated as identity achievement and moratorium behave similarly, and the men and women of foreclosure and identity-diffused status constitute another grouping on a need-for-achievement dimension. The identity achievement and moratorium groups had the highest scores. However, gender differences were found for the "fear of success" dimension. Identity achievement and moratorium men had *low* fear of success. Yet the women of the same identity types expressed a *high* fear of success. Orlofsky interprets these results as reflecting the greater conflict experienced by women who pursue nontraditional, more achievement-oriented goals. It is conceivable that such conflicts will diminish as the role of women in our society changes. Further studies will have to address the function of the foreclosure status for women. The foreclosure status may become less adaptive as society becomes more supportive of innovative, nontraditional roles for women, requiring them to establish the autonomy of a fully developed identity.[3]

2.2. Ego Processes and Personality Formation

2.2.1 Block's Conceputalization: Ego and Personality Development

In contrast to the preceding approaches that emphasize identity configurations of adolescent formation, Block conceputalizes specific organizing mechanisms in personality development. Through longitudinal studies, Block

---

[3]Waterman (1982) summarizes current research, suggesting that men and women undergo similar patterns of identity development, but the identity statuses have different meanings for men and for women.

amasses considerable evidence for the time-patterned continuity of trait clusters in individuals (J. Block, 1971, 1981). Block views personality development as the changing capacity of the ego to mediate between internal drives and external opportunity, in order to foster expression of each person's unique individuality (J. Block, 1971).

In Block's perspective, two basic constructs are "ego control" and "ego resiliency." Block defines *ego control* as "the threshold of operating characteristic of an individual with regard to the expression or containment of impulses, feelings and desires" (J. Block & Block, 1977, p. 2). Block's definition is derived from the idea that the common denominator of the ego functions is the control and modulation of impulse. Ego control refers to characteristic styles of controlling impulses, feelings, and desires.

The other construct, *ego resiliency*, refers to the person's elasticity: "the dynamic capacity of the individual to modify his modal level of ego control, in either direction, as a function of the demand characteristic of the environmental context" (J. Block & Block, 1977, p. 2). In other words, ego resiliency refers to the person's ability to adapt flexibility to changing circumstances.

Together, ego resiliency and ego control are considered condensed abstractions that contain all the essential qualities of the analytic concept of ego (J.H. Block & Block, 1979). These constructs are currently being directly assessed in a longitudinal study (J. Block & Block, 1977; J. H. Block & Block, 1979). Characteristics observed in the young children of the study illustrate the meanings of ego resiliency and control. For example, 3- and 4-year-old children scoring high on measures of ego control are described as more active, assertive, aggressive, competitive, and private than low scorers. Those scoring high on the measures of ego resiliency are described as more empathic, able to cope with stress, emotionally appropriate, self-reliant, and novelty seeking (J. H. Block, & Block, 1979). Future assessments planned for these children in their adolescent years are designed to generate information on specific role of ego resiliency and ego control in adolescent personality development.

An earlier study conducted by Block (1971) explored the role of the ego in the personality development of adolescents and adults. Although Block does not specifically assess ego resiliency and ego control in this analysis, he does view ego structures as major aspects of personality. The permeability and flexibility of the ego determines the way the individual responds to inner motivations and external pressures, providing the basis for each personality to develop. Block derives a number of personality types from his observations and suggests a connection between personality and ego functioning. The relationship will become clear as we describe the types of personality development and the approach of the original investigation.

Block's sample included participants from two longitudinal studies conducted by MacFarlane (1938) and Jones (1938) of the Institute for Human Development. These subjects had been frequently and intensively interviewed from infancy through high school. In addition, Block contacted 171 subjects in adulthood and conducted interviews with them. They ranged in age from 30 to 37 at the time of their adult interviews. To carry out the analyses, each subject's voluminous data set were divided into three distinct periods: early adolescence (junior high school), middle adolescence (senior high school), and adulthood. Dossiers were prepared consisting of interview, testing, and observational reports, as well as demographic and physical characteristics. Judges then evaluated the subject's personality characteristics at each period. The rating procedure used was Block's Q-Sort, which specified rules for scaling the group of personality characteristics. These personality ratings led to quantified (in terms of salience) descriptions of each subject's personality features ("traits"). Of major importance is the fact that the Q-Sort provided a way of reliably assessing personality change over time, both within the individual and in comparison to others.

Through these analyses Block delineates major groupings of men and women with similar constellations of traits. In addition, he argues that certain characteristics found in adolescence are predictive of later traits and behavior, and can be related to various types of ego control. Those personality types with marked undercontrol occupying one end of the continuum. Those for whom overcontrol is the dominant feature of their character structure are found at the opposite end. The place of ego control in the various types of personality development is illustrated in the following descriptions. Included are comparisons of data from adolescence and adulthood. Although Block distinguishes between early and middle adolescence, we combine these periods for our discussion.

2.2.2. Measurement and Application

2.2.2a. Types of Personality Development in Males. The first group identified by Block is called *ego resilients.* These are males who may seem to have an appropriate degree of ego resiliency and control. During both high school and adulthood, these men were rated as being ambitious, bright, satisfied-with-self, and likeable. Cheerful, poised, and leaders in youth, these individuals became productive, self-confident, maritally stable, and economically successful. "In his surgent maturity, the ego resilient male accepts himself and has a sense of worth because he can see his personal value confirmed by family and friends and society" (J. Block, 1971, p. 150). In terms of identity formation, these individuals seem to have developed clear, effective identities easily and naturally—without apparent turmoil or upheaval. Block cites this as evidence that personal crisis and self-confrontation may not be necessary for identity formation (J. Block, 1971).

The unsettled undercontrollers stand in marked contrast to the ego resilients and are characterized as lacking ego control; they are impulsive, hostile, rebellious, and self-dramatizing. A consistent thread from adolescence through adulthood is their quest for meaning and self-definition. Although intellectually and socially the most advanced of the sample, these males have the most difficulty achieving an identity. Thus, their relationships as adults are not yet mature and they have not yet found a stable place in society. Not overtly maladjusted, and clearly intelligent, interesting, and likeable, they seem to fall short of potential in their pervasive impulsivity and lack of self-definition.

The next group, *vulnerable overcontrollers,* occupies the pole of the ego control continuum furthest from the unsettled undercontrollers. Extremely constricted in high school, the overcontrollers remain withdrawn, submissive, ruminative, aloof, and uncomfortable with role changes. Ego defense mechanisms developed in their adolescence include denial, delaying, and simplification. Although these defenses served to allay the anxieties of youth, they are not adequate to adult tasks. Thus, as adults, these males are left with rigidified ego structures that do not allow them to cope flexibly with their situations in life. Described as unhappy and not knowing how to live, they do not form solid identities as adolescents, nor do they develop satisfying vocations or relationships. Their excessive constriction keeps them disengaged and set somewhat apart from the mainstream of life.

The *anomic extroverts* display a thin veneer of conformity that masks a sense of valuelessness and an absence of inner life. They change much more than the other groups—from gregarious, assertive, peer-oriented, and undercontrolled youth to hostile, fearful, aloof, and overcontrolled adults. A repressive character structure and anomie in adulthood maintains an emptiness that exists in place of a firm identity and the accompanying personal adequacy and goal orientation.

After a troublesome adolescence, the *belated adjusters* later achieve competence and self-definition. In youth, they were negativistic, easily frustrated, nonverbal, and limit testing. As adults, they abandon their defensive, rebellious behavior and submit to circumscribed and easily predictable lives. Block speculates that with their limited talents and somewhat lower socioeconomic origins, these men were able to achieve stable personality development by adopting the roles of steady workers and husbands and fathers.

In these types of development, we see various manifestations of ego control leading to various patterns of behavior. Block maintains that the ability to delay gratification and modulate impulse is a central determinant of personality, and he illustrates this in the types just described. Implications are apparent for the males: overcontrol leads to constricted behavior, shallow relationships, and a disengagement with the mainstream of life. Undercontrol is also problematic, leading to impulsivity that prevents the type of commitments necessary to form a stable sense of self. Appropriate control is associated with solid relationships, firm self-definition, and flexible interaction with others and with life's challenges.

Such clear patterns are more difficult to find in the women of this sample. As we consider the various types of female development, it will be seen that a single, extremely overcontrolling group does not exist. Block speculates that this is because

> the cultural setting within which the female subjects developed set great store on conventions by which a girl should live (or at least disguise her spontaneities). . . . It may not be unfair to suggest that overcontrol for our girls and women is appropriate control, given the environmental structuring and values around them. (J. Block, 1971, p. 265)

This point is illustrated in the following descriptions.

2.2.2b. Types of Personality Development in Females. The *female prototypes*, as the name implies, are individuals who have manifested the qualities deemed as desirable and appropriate for women in our culture. In high school, these adolescent girls were considered likeable, warm, cheerful, attractive, poised, and aware of their social stimulus value. Well-integrated into the peer culture, they were gregarious, oriented toward boys, talkative, and social. As adults, these women are considered at ease with themselves, nurturant, attractive, and good communicators of social values. The mad social whirl and period of undercontrol during adolescence is replaced with overcontrol as they become fastidious, poised, but somewhat clichéd women. Self-centeredness and undercontrol characterized the early years, whereas self-control and concern with instilling proper virtues into others are the dominant aspects of these females' later lives. They form the stable base of family and community and are often seen as indispensable for their contributions.

In contrast, the core characteristics of the *vulnerable undercontrollers* are a combination of impulsivity of action and a submissiveness that leaves them open to exploitation by others. In adolescence and adulthood, they are undercontrolled, self-indulgent, self-dramatizing, changeable, dependent, and unpredictable. They marry earlier and more often and seem unable to control their environments or to order their lives. The impulsivity, disorganization, and dependency that characterize these women remained throughout the lifespan studied.

In addition to the vulnerable undercontrollers, the group termed *dominating narcissists* share the central characteristics of impulsivity and minimal ego control. The name appropriately indicates the self-absorption and aggression displayed by these women. In adolescence and adulthood, they are condescending, deceitful, self-indulgent, and tend to push limits. As adolescents, they were irritable, distrustful, rebellious, undercontrolled, and defensive. As adults, they are self-satisfied, power-oriented, poised, straightforward, and likely to express hostility directly. Thus, the pseudosophisticated, fickle, suspicious youth become productive, socially adept, decisive, manipulative adults.

Hyperfeminine repressives were described in similar ways to the traditional hysteric: fitful emotionality alternating with blandness, and sexuality that is both naive and deliberate. In high school and adulthood, they are feminine, repressive, and withdrawing, concerned with their bodies and physical appearance, dependent, and fearful. This submissive, hyperconventional, sexually naive but provocative adolescent is rated as an aloof, but erotic, brittle, inflexible adult who constantly feels victimized.

The last two groups are both closer to the "overcontrol" end of the ego control continuum, and are distinct from the three other female groups. Although similar to the female prototypes in their tendency toward control, the paths of development for the *cognitive copers* and the *lonely independents* are very different, and do not evidence the same degree of continuity in personality development. For example, the cognitive copers appeared somewhat inadequate as adolescents, but become competent as adults, progressing from average intellectual and socioeconomic origins to high degrees of education and upward mobility. Overcontrolled, independent, aloof, intellectualized, and overly introspective, the adolescent in this grouping was also self-confident, well-organized, and effective. As an adult, she is rated as warmer, more relaxed, and less supercilious. She is described as effective, guided by goals and principles, calm, valuing intellectual matters and her independence.

Finally, the *lonely independents* are the women whose core characteristics are assertiveness, autonomy, and interpersonal distance. As youth, they were ambitious, critical, interesting, independent, intellectual, rebellious, and in search of meaning. Their intellectual capacity remains in adulthood as well as their independence, objectivity, and lack of personal meaning in life. Yet these women tend to move farther from interpersonal closeness as time goes on, becoming increasingly independent and intellectually active. Their early rebellious, gregarious characteristics fade and are replaced by rigid independence and self-control.

Through these empirically generated types, Block argues that impulse control is a basic determinant of personality. He further emphasizes both the continuity and discontinuity of personality formation and the importance of maintaining a developmental perspective. By relating the various character types to an underlying continuum, he maintains the conceptual clarity of a typology as well as the cohesion of a longitudinal perspective. In our next section, we consider another investigator who conceptualizes personality types within a developmental continuum.

2.3. Ego Development

2.3.1 Loevinger's Model: Stages of Ego Development

Loevinger's view of ego development differs from concepts we have discussed and adds an important dimension to our understanding of adolescence. In Loevinger's usage, ego development refers to stages and includes what others have separately called "moralization," "integration," "relatability," "self-system," and "cognitive complexity" (Adler, in Ansbacher & Rowena, 1956; Harvey, Hunt, & Schroeder, 1961; Kohlberg, 1964; Piaget, 1932; H. S. Sullivan, 1953; C. Sullivan, Grant, & Grant, 1957). Her conception is derived from theories that focus on frameworks of meaning and an assumption of sequential stages. Loevinger depicts ego development as the formation of increasingly complex views of oneself in relation to the world: continual changes in one's framework of meaning through the achievement of successive stages of development. This is a different picture of adolescent development than that provided by emphasis on personality type (Block) or identity status (Marcia).

Loevinger's formulation focuses on the adolescent's *own* perspective. Her view is summarized in the statement that the search for meaning is not something the ego *does* but what the ego *is* (Fingarette, 1963). Thus, the ego is basically a search for coherent meaning, and ego development is the change that occurs during this search. The complexity of this conception is best captured by describing the successive stages (Loevinger, 1966, 1973; Loevinger & Wessler, 1970). Each stage is considered a qualitatively unique milestone in the sequence of development. A continuum is created that embodies sequential changes in structures of meaning and of character. The stages and associated character types form patterns that can be operationally defined.

As mentioned earlier, one of the important qualities of Loevinger's construct is that it is measurable and thereby amenable to empirical investigation. Ego development is measured through a specifically designed sentence completion test (SCT) with separate forms for men, women, boys, and girls. The comprehensive scoring system, rationale, and measurement properties have been carefully studied, and strong evidence for validity and reliability has been presented (Hauser, 1976a; Holt, 1974; Loevinger, 1979; Loevinger & Wessler, 1970; Loevinger, Wessler, & Redmore, 1980; C. Redmore and Waldman, 1975).

2.3.2. Description of the Stages

Loevinger describes a series of sequentially ordered ego development stages. Although they may be correlated with chronological age, they are defined "independently" of age and are seen as forming an invariant hierarchical order. Each stage is more complex than the last and none can be skipped in the course of development. However, different individuals may not develop beyond various stages. Among adults there are representatives of each stage, who can then be characterized in terms of the features specific to the stage at which they have stopped. Consequently, in addition to being an invariant sequence of stages, Loevinger's framework presents a typology of individual differences in the form of "character styles," with some parallel to Block's personality types.

The first stage consists of presocial and symbiotic phases. In the *presocial* phase, the infant is described as oblivious to all but gratification of his immediate needs. At this point, animate and inanimate parts of the environment are not distinguished. In the *symbiotic* phase, however, the child has a strong attachment to the mother (or mother surrogate), distinguishing this figure from the rest of his environment. But he does not yet differentiate himself from his mother. This first stage comes to an end at approximately the time that use of language is acquired. Hence, this earliest period is inaccessible to study by means of techniques (such as the sentence completion test), which rely on verbal language.

The second stage is the *impulsive* one. Impulses are the predominant factors in the individual's life; their control is undependable; exercise of one's own will is desirable. Rules are not recognized as such. Actions are seen as "bad" or "good" because they are either punished or rewarded. Conscious concerns are with the satisfaction of physical needs, including sexual and aggressive wishes. The view of the world is egocentric and concrete. This is the first stage tapped by the sentence-completion test.

The third stage is the *self-protective* one. The child understands there are rules, but obeys them only out of self-interest. Morality is purely pragmatic: Avoid what is bad, for you will be caught. Interpersonal relations are manipulative and exploitative. Independence grows during this stage: children doing things for themselves; adolescents feeling they don't need "the older generation." Conscious concerns are with control and domination.

Examples of responses at this level are:

- *What gets me into trouble is* "I can't tell a fib without smiling."
- *Most men think that women* "are out for what they can get."
- *When people are helpless* "I don't like to be bothered with them."

The next stage is a transitional one between the self-protective and conformist stages. As with the other transitional stages, it was derived through empirical work with the SCT and the theoretical basis is unclear (Hoppe, 1972). Responses are neither low enough to merit an opportunistic rating nor complex enough to warrant a higher rating. Taken together, these first three stages are often regarded as representing an early, or "preconformist," level of development.

The fourth stage, the *conformist* one, is reached by most people sometime during childhood or adolescence (Loevinger & Wessler, 1970). Rules are obeyed at this stage simply because they are rules; morality is only partly internalized. Shame for transgression now becomes meaningful. Some interpersonal reciprocity begins to occur, although relations are viewed in terms of concrete events and actions rather than feelings or motives. Where inner states are expressed, they are stereotyped, clichéd, and moralistic. Conscious concern is with material things, appearance, reputation, and status. Examples of conformist-level responses are:

- *What gets me into trouble is* "talking."
- *The thing I like about myself is* "my morals."
- *When people are helpless* "I try to help them."

The transition between *conformist* and *conscientious* stages is called the *self aware* because of the growing realization that—the "right ways" of living are relative to situations. Contingencies become recognized, and although the world is still viewed in rather global, stereotyped terms, there is somewhat less absolution than in previous stages. Self-awareness grows, as do capacities for self-criticism, introspection, and psychological causation. With the growing self-awareness comes some freedom from dependence on the social group for strict behavioral guidelines. Various studies report more persons at this stage than at any other (Haan, Stroud, & Holstein, 1973; Harakel, 1971; Lambert, 1972; C. Redmore & Waldman, 1975).

The fifth stage, *conscientious*, marks the consolidation of the changes of the transitional period. Moral standards are internalized; interpersonal relations are seen in terms of feelings rather than of actions, and individual differences rather than rigid stereotypes are perceived. Inner standards become the central guideline for behavior, surpassing external rules and recognition as the primary motivator. The capacity for self-criticism characterizes this stage, and conscious thoughts focus on ideals, achievements, obligations, and traits. Responses at the conscientious level include:

- *My main problem is* "lack of time to accomplish everything."
- *A woman feels good when* "she has accomplished something to help other people."

- *When people are helpless* "I sympathize, unless they are unwilling to help themselves."

The next stage, called *individualistic*, is the third transitional one. People at this stage have more complex responses than those at earlier levels. People show an ability to tolerate paradoxical relationships and no longer attempt to reduce them to polar opposites (Hoppe, 1972). Characteristic of this stage is the value placed on interpersonal relationships. A growing recognition of emotional interdependence accompanies decrease in emphasis on striving on independence. This contrasts noticeably with the emphasis on ideals and achievements seen at the *conformist* level.

The sixth stage is labeled *autonomous*. The central issue is coping with inner conflicts, conflicts between duties and needs, and conflicting ideals. Such conflicts were present at earlier stages, but were not faced and coped with as directly as in this period. Along with facing and tolerating the complexities of one's own inner conflict comes a tolerance of other solutions. The moral condemnation of earlier stages is absent. The characteristic of this stage is seen in the interpersonal realm: Although at the *individualistic* level interdependence is recognized, at this next level the need for autonomy is also deemed important. The person at this stage sees the need for others to learn from their own mistakes, rather than feeling obligated to prevent others from erring. Conscious concerns are with self-fulfillment, individuality, role-differentiation, and complexity of options. Typical responses are:

- *A woman should always* "strike a balance between her own wants and satisfactions and those of her family."
- *What gets me into trouble is* "expecting too much of others."
- *When people are helpless* "it is best to aid them to help themselves than to prolong their helplessness and dependency on others."

The highest stage is the *integrated* level. At this level, the person is "proceeding beyond coping with conflicts to reconciliation of conflicting demands, and, where necessary, renunciation of the unattainable" (Loevinger, 1966, p. 200). The highest level is more theoretical than empirical and one finds no more than 1% of persons at this stage (Loevinger & Wessler, 1970, p. 9). At this level, all responses are unique. They also tend to combine at least two aspects in a single response: usually commenting on interpersonal relations, rules, standards or conflicts, and/or conceptual complexity (Loevinger, 1966). It is interesting to note that Loevinger states that none of her high-level respondents were professionals and none had had training in writing; yet all responses were touching or poetic. It is difficult to combat the tendency to idealize this stage as the Utopian height of their development, but such difficulties pervade most developmental hierarchies. Some examples of this stage are:

- *A woman feels good when* "her sex life goes well and she and her husband are 'sympatico'."
- *A good mother* "is kind, consistent, tender, sensitive, and always aware a child is master of its own soul."
- *The worst thing about being a woman* "is accepting your position as a woman and an individual, but once found ceases to be the worst and becomes the best."

To summarize, this model of ego development conceptualizes seven sequential invariantly ordered stages. Since the seven stages were originally described, three intermediate or transitional ones have been added on the basis of additional data and theoretical considerations. It is possible, indeed probable, that the sequence can be interrupted at any point in development. The consequence of this interruption is a "character style" corresponding to the features of the particular stage in which progression ceased. Table 1 lists characteristics of the seven original stages, together with three transitional stages.[4]

2.3.3. Measurement and Application

Loevinger conceives of ego development as progressive changes in the way one views the world. It is appropriate, then, that measurement of this development is a sentence-completion test, since this allows the individual to project his or her framework of meaning into the testing stimulus. Responses are scored with the use of a complex scoring system, elaborated in a two-volume manual (Loevinger & Wessler, 1970; Loevinger, Wessler, & Redmore, 1980). A set of graduated exercises permits investigators to become reliable scorers, consistent with those trained by Loevinger's group (Loevinger & Wessler, 1970). High interrater reliability has been reported in a number of studies, with correlations ranging from 61% to 92% (Cox, 1974; Hoppe, 1972; Loevinger & Wessler, 1970).

The assumption of the scoring paradigm is that each person has a core level of ego functioning. This core functioning is determined by assigning an ego development level to a person that is based on the scores of the sentence-completion test. Each response is scored by assigning it to one of nine levels,

---

[4]These quotes are taken from individual responses given in Loevinger and Wessler, 1970 and Hauser, 1976a.

which are based on the specific characteristics of the successive stages previously summarized. A total overall score is then derived from the individual
stem scores and is considered the person's ego development level.[5]

The test has been shown to be "carefully constructed and standardized
in terms of its form, administration and scoring procedures" (Hauser, 1976,
p. 951). High reliability demonstrated for both the test and the scoring system
encourage confidence in Loevinger's model. Additional studies using a variety
of methods and designs also offer much evidence for several forms of validity
of the ego development test. Comprehensive reviews can be found in Hauser
(1976a) and Loevinger (1979). Here, we consider only those topics that are
relevant to adolescent development.

2.3.3a. IQ and Verbal Fluency. The first topic concerns the relationship
of IQ and verbal fluency to ego development: Do these factors affect performance on the test? Or, at a greater extreme, is the measure of ego development just one more assessment of intelligence and verbal fluency? By
clarifying the relationship of these variables, we can better integrate them
into our understanding of adolescent development.

Various studies suggest that IQ and ego level are related, yet not
completely overlapping dimensions. Ego development has been found to be
significantly related to IQ and to measures of verbal fluency and quantitative
ability (Bonneville, 1978; Hoppe, 1972; Redmore & Loevinger, 1979). In
part, greater verbosity and higher IQ is expected to be related to higher
levels of ego development as

> conceptual complexity is part of what we call ego development. To exhibit con-
> ceptual complexity in a manner recognizable to raters requires (usually) longer
> responses, combining several ideas. (Loevinger & Wessler, 1970, p. 51)

It is possible therefore, that IQ is a confounding variable. Equally
possible is the alternative that ego level contaminates IQ. As Loevinger
(1979) points out, many schools have noted uncooperative adolescents (presumably of low ego level) who seemed stupid but actually were not. Yet
another alternative is that ego level and IQ are correlated; possibly in varying
degrees at different levels (Hauser, 1976a).

2.3.3b. Behavioral Correlates. Another important relationship is that
between ego level and actual behavior. Loevinger's model suggests that we
should expect to find links between stages of ego development and patterns
of behavior. For example, those of the conformist level are more likely to
behave in a stereotyped, socially predictable manner. Those at a lower level
would yield more readily to impulse and exploitation of others (see Table
1).

---

[5]For a thorough discussion of the scoring methods that have been developed for the SCT, see
Hauser (1976a) and Loevinger and Wessler (1970).

TABLE 1. Milestones of Ego Development[a]

Stage	Impulse control, "moral" style	Interpersonal style	Conscious preoccupations	Cognitive style
Presocial (I-1)		Autistic	Self vs. nonself	
Symbiotic (I-1)		Symbiotic	Self vs. nonself	
Impulsive (I-2)	Impulsive, fear	Receiving, dependent, exploitive	Bodily feelings, especially sexual and aggressive	Stereotypy, conceptual confusion
Self-protective (Delta)	Fear of being caught, externalizing blame, opportunistic	Wary, manipulative, exploitive	Self-protection, wishes, things, advantages, control	
Transition from self-protective to conformist (Delta/3)	Obedience and conformity to social norms are simple and absolute rules	Manipulative, obedient	Concrete aspects of traditional sex roles, physical causation as opposed to phychological causation	Conceptual simplicity, stereotypes
Conformist (I-3)	Conformity to external rules, shame, guilt for breaking rules	Belonging, helping, superficial niceness	Appearance, social acceptability, banal feelings, behavior	Conceptual simplicity, stereotypes, clichés
Transition from conformist to conscientious; self-consciousness (I-3/4)	Dawning realization of standards, contingencies, self-criticism	Being helpful, deepened interest in interpersonal relations	Consciousness of the self as separate from the group, recognition of psychological causation	Awareness of individual differences in attitudes, interests and abilities; mentioned in global and broad terms
Conscientious (I-4)	Self-evaluated standards, self-criticism	Intensive, responsible, mutual, concern for communication	Differentiated feelings, motives for behavior, self-respect, achievements, traits, expression	Conceptual complexity, idea of patterning

Stage	Impulse control, "moral" style	Interpersonal style	Conscious preoccupations	Cognitive style
Transition from conscientious to autonomous	Individuality, coping with inner conflict	Cherishing of interpersonal relations	Communicating, expressing ideas and feelings, process and change	Toleration for paradox and contradiction
Autonomous (I-5)	Add: Coping with conflicting inner needs[b]	Add: Respect for autonomy	Vividly conveyed feelings, integration of physiological and psychological causation of behavior, role conception, self-fulfillment, self in social context	Increased conceptual complexity; complex patterns, toleration for ambiguity, broad scope, objectivity
Integrated (I-6)	Add: Reconciling inner conflicts, renunciation of unattainable[b]	Add: Cherishing of individuality	Add: Identity	

[a] Adapted by Hauser, 1976a, from Loevinger and Wessler, 1970.
[b] "Add" means in addition to the description applying to the previous level.

Investigators have chosen different methods to address this issue. Behavioral patterns in school children and juvenile offenders have been examined (Blasi, 1971; Cox, 1974; S. Frank & Quinlan, 1976; Hoppe, 1972). Vocational choice (Schain, 1974; Sheridan, 1974) and personal growth (Goldberger, 1977) in college students have been studied. These investigations have identified some clear behavior patterns that correlate with ego level. In general, preconformists are found to be rebellious, immature, impulsive, and often badly behaved (Blasi, 1971, 1976; S. Frank & Quinlan, 1976). Prison inmates at the impulsive stage are described as impulsive, easily angered, and lacking discipline (S. Frank & Quinlan, 1976). Delinquent adolescents at the impulsive and self-protective stages report a significantly higher number of street fights, homosexual relations, and attempts to run away than from all other stages (S. Frank & Quinlan, 1976). Among black sixth-grade boys and girls, Blasi (1971, 1976) found that children at the impulsive stage were *rated lower* intellectually than other children. They also had short attention spans, were restless, misbehaved, and performed poorly in discrimination of feelings. Although they could see themselves as the cause for bad behavior, they tried to shift blame and were preoccupied with "ratting on one another."

Individuals at the self-protective stage were somewhat more cooperative and gregarious (Blasi, 1971, 1976; S. Frank & Quinlan, 1976). For example, the prisoners were seen as harshly critical of procedures. Self-protective stage children in Blasi's study became angry and defiant when caught or shamed, tending to role-play sneaky, opportunistic heroes who demanded leniency when caught.

Children and adults at the conformist levels are more rule oriented, better adjusted, and not rebellious. The inmates were characterized as dogmatic (S. Frank & Quinlin, 1976); whereas conformist children emphasized obedience, interpreted rules literally, and seemed motivated by a love for and loyalty to authority (Blasi, 1971). These behavior patterns parallel Loevinger's description of conformist stage features, emphasizing social approval, authority, and rules. In general, the studies show that youth and adults at lower levels of ego development tend to be less disciplined, more impulsive, opportunistic, and self-serving.[6]

A consistent finding is that both juvenile and adult offenders are at lower levels of ego development than their age-peers. An intriguing question, then, is whether treatment aimed specifically at enhancing ego development would be helpful for these individuals. It may be that advances in ego level would facilitate increased behavioral control and rule-orientation qualities lacking in the lower stages of ego development. Some studies suggest that it

[6]Individuals at higher levels are more reliant on rules and society's approval and are more obedient and cooperative.

is possible to demonstrate gains in ego level after psychological education (P. J. Sullivan, 1975). What is not clear is whether these gains are accompanied by behavior change. Also unknown is the relationship between ego level and deviant behaviors such as forms of delinquency and psychopathology. This relationship must be more clearly articulated before predictions can be made regarding how enhanced ego development might influence behavior change.

If consistent relationships between ego level and behavior are shown to exist, clinical insight about adolescents could be greatly increased, as would the repertoire of possible behavioral interventions. For example, an adolescent at low ego level may display impulsive behavior. From what we know of impulsive stage outlooks, we would suspect that a behavior strategy could be effectively applied in his home, classroom, or treatment setting with clear rewards for desired goals. Adolescents at higher levels of ego development may be expected to be more responsive to authority. Their behavior can probably be affected through clear delineation of rules and contingencies for social approval. These are but two examples; a great many possibilities exist.

The optimal way to undertake studies of the link between ego development level and action seems to be by looking for behaviors generated by the interplay between certain situations and specific ego development levels. For example, Cox (1974) investigated helping behavior in white and black eighth-grade middle-class boys. He found that low ego subjects helped the lowest percentage of time in a low-help condition and the highest percentage of time in a high-help condition. Neither ego level nor condition of prior help alone correlated significantly with helping behavior. This illustrates the usefulness of making predictions specifically derived from the theoretically described characteristics of each stage *and* allowing for situation dependent change.

Loevinger's model of ego development offers many possibilities for enriching our knowledge of adolescence. It is clear that her ego development measure is not simply another label for intelligence or verbal ability. Yet these dimensions are nonetheless related ones. In addition, there is strong evidence for connections between behavioral patterns and ego stages. Although there is room for further empirical research in such areas as ego development and adolescent medical illness (Hauser, Pollets, Turner, Jacobson, Powers, & Noam, 1979; Hauser et al., 1982) and psychiatric illness (Gold, 1980), we can already envision numerous implications for approaches to working with normal or ill adolescents.

2.4. Self-images in Adolescent Development

Even the most casual observer of adolescents cannot fail to be impressed with their heightened self-focus and frequent self-absorption, in tandem with the "imaginary audience" (Elkind, 1967), which is watching and so ready to accept or reject on the basis of the right clothes or lingo. These vivid descriptions by family members as well as many clinicians (Erikson, 1968; Wheelis, 1958) are associated with thorny theoretical and empirical questions regarding the relationship of the self to such constructs as ego identity and ego development. In this section we consider these questions in light of recent conceptual and empirical advances. The questions go further than an academic interest in clarifying theoretical issues. Clinicians depicting medically ill adolescents often reflect confusions about these developmental and personality dimensions as they speak of Jack's "self-esteem," when alluding to his ego development, or Lynn's "self-image" when referring to her low self-esteem. Our conceptual and empirical review is intended to lead to greater clarity and subsequent more precise application of these important yet elusive terms.

2.4.1. The Self in Adolescence: Prefix and Construct

Interest in aspects of the self reaches across many disciplines—developmental psychology, life-span studies, social psychology, personality theory, and psychoanalysis. Yet as Harter (1984) notes in her superb review of this area, "the 'self' functions more as a prefix than as a legitimate construct in its own right" (p. 1). Among the terms that use self in this way are *self-disclosure, self-concept, self-image, self-theory, self-esteem, self-awareness,* and *self-criticism*. Some of these concepts are most relevant to the complexities of identity and ego development. In later passages we review such linkages. But before dealing with the prefix terms, we must take up a broader theme, that of the self as a *construct*.

An important distinction was first articulated by William James in 1890, as "he contrasted *two fundamental aspects* of the self, the self as actor or *subject*, the 'I', and the self as an object of one's knowledge and evaluation, the 'Me' " (Harter, 1984, p. 2). The "I" is the active observer, the *constructor* of the self, whereas the "Me" is *generated by* the observing process when its attention is turned toward the self. Despite the recurrence of this theme in theoretical discussions, most of the empirical adolescent studies in this area have attended to the "Me," the self as object of one's knowledge and evaluation (Harter, 1984; Wylie, 1974, 1978). When we consider the "I" we encounter fascinating questions that ask about the processes underlying and influencing self-knowledge and the structure of self-understandings. Studies of the "Me" refer to the content of self-experience—the nature of self-esteem, self-awareness, self-control, and self-reported ideology. Our own studies (Hauser, 1971, 1976b,c) have focused upon the "I," exploring certain structures of self-experience (self-images) and their relation to identity development. Contrasting with Marcia's emphasis on more conscious contents (e.g., occupational ideology) of adolescent thought, our analyses are on less conscious self-image structures or processes.

Discussion of both ego and identity development refers to self-images directly and implicitly (Erikson, 1959; Loevinger, 1976; Loevinger & Wessler, 1970). Within many of Erikson's writings about identity development are descriptions of self-image processes and predictions about them (Harter, 1984; Hauser, 1971). Using these many references to self-images, we developed a series of operational definitions of identity variants and applied them in a study of black and of white adolescents (Hauser, 1984). These definitions were based on specific combinations of self-image processes (integration and continuity), which could be empirically assessed. Our reasoning, drawn from clinical observations and theory, was that these were the self-processes that were fundamental to the establishment of the adolescent's sense of ego identity.[7]

An adolescent (or adult, for that matter) holds *multiple* self-images, variously organized and structured, and at varying levels of awareness. In other words we cannot refer to *an* adolescent self-image, or global self-esteem, as a state that can be assessed in a single testing. The picture is a more complex one. The adolescent self can be likened to a theory constructed to organize one's thinking about oneself and one's relationship to the social world. Several proponents of this orientation have emphasized the "social world" facets of the "theory" (Brim, 1976; Kelly, 1955; Sarbin, 1962). Other recent discussions have included additional facets of self-theory through the application of information-processing models (Markus, 1977) or the attention or self-regulation functions of self-theories (Lynch, 1981). The most compelling formulation is presented by Epstein (1973):

> I submit that the self concept is a self-theory. It is a theory that the individual has unwittingly constructed about himself as an experiencing, functioning individual, and is a part of a broader theory which he holds with respect to the entire range of significant experience . . . there are major postulate systems for the nature of the world, for the nature of the self, and for their interaction. (p. 407)

The structure of this self-theory, the system of self-conceptions, reflects the nature of the individual's identity formation. We are referring to the adolescent "I," the constructor of self-knowledge of experience. Information regarding his or her "postulate systems" (or basic assumptions) pertains to the "Me," the contents of the self-theory. In our studies we define self-images

[7]It is of interest that Harter (1984) includes these two components together with a third, in her recent review of identity and self: "For Erikson, the task of this period (adolescence) is the establishment of a sense of ego identity, which involves three components: a sense of *unity* among one's self-conceptions, a sense of *continuity* of the self-attributes over time, and a sense of the mutuality between the individual's conception of the self and those which significant others hold of him/her" (p. 72). Although our formulations do not include this third dimension, it is an important one, worthy of consideration in future empirical studies.

as consisting of those concepts at varying levels of awareness, by which an individual characterizes himself. *Identity* has been described as referring to a "consolidation" or "integration" of multiple components, arrays of ego functions and identifications. The term *self-image* includes these other components; self-image is at a higher level of abstraction and can be analyzed into these subunits. There are many techniques available to determine the array and nature of an individual's self-images. Psychoanalysis and psychotherapy represent one approach. A more immediately objective method is the *Q*-Sort, in which the individual considers and sorts self-descriptive statements. This is the method we have used to study the individual's self-images.

The instrument we use to tap self-images is a specially designed and administered *Q*-sort. In contrast to other studies (J. Block, 1961; Engel, 1959) that use standardized decks of self-descriptive cards and a forced distribution (Stephenson, 1953), we request our subjects to sort a deck of "I" statements, which he or she has expressed during clinical research interviews (Hauser, 1971; Hauser, Jacobson, Noam, & Powers, 1983), or small group discussions (Hauser, 1976c). From the tapes of these interviews or meetings a separate *Q*-Sort deck is generated for each adolescent in our studies.[8] The deck consists of multiple cards, each containing an "I" statement that had been expressed by the adolescent and was about his or her thoughts or feelings about outside events or people as well as self-attributes. Such statements include self-descriptions, personal positions, and self-evaluations.

An adolescent is asked to sort these statements in terms of several self-images, each from a different perspective. For example, during one sort, the subject is asked to rate each card according to whether it is something he would "say, think, or feel." During another sort he is asked to rate the identical statements, imagining himself "to be a perfect son in the eyes of his mother." He rates the statements accordingly: Are these ideas he would "say, think, or feel" if he were that person?

2.4.2. Identity Formation and Identity Variants through Self-images

In one of his early theoretical papers Erikson (1959) described identity formation as consisting of "an evolving configuration of constitutional givens, idiosyncratic libidinal needs, favored capacities, significant identifications, effective defenses, successful sublimations, and consistent roles" (p. 116).

---

[8]Our investigations have used self-descriptive statements culled from two settings—individual interviews and group discussions. In summarizing the various studies, we will indicate the particular setting. In general, the individual interview has been the more consistently used source of statements.

Although this is obviously a comprehensive definition, it is hardly operationally suitable for empirical research. Its use in the design of a nonclinical study of identity development requires considerable transformation. Our definition of identity formation, and its variants, was based on the following operational statement: Identity formation refers to specific processes of an individual's self-images, namely (a) their structural integration and (b) their temporal stability. Changes in either or both of these processes are indicative of specific types of identity formation.

Self-image was defined in a preceding section. The undefined terms of the definition, then, are "structural integration" and "temporal stability." The concept of *structural integration* refers to how one's self-images are organized at a specific point in time. The many descriptions by Erikson and others (Strauss, 1959; Wheelis, 1958) emphasize the "synthesis" and "consolidation" aspects of identity formation. They note the growing organic whole, the coherence of elements. Structural integration is the translation of this observation to a measurable function. The network of an individual's self-images at any moment represents their structural integration. The interrelations of the self-images comprising this network are reflected in their correlations with one another. We derive an index of structural integration by taking the average of all of an individual's self-image correlations at a particular testing. For example, if the subject is sorting seven self-images, his structural integration score would be the average of the matrix of 21 correlations among these self-images. Changes in this process are discovered by retesting at a later time using the same self-image stimuli.

In discussing identity formation, writers often refer to *continuity* and stabilization. One important meaning of these terms is constancy of self-images, the degree to which present self-concepts resemble those of the past. In clinical discussions, an "increasing sense of continuity" is often seen as one important sign of identity development. The assessment of the continuity of self as self-image is accomplished through comparing the two different arrays of self-image statements the subject uses to express the image at two separate points in time. The quantitative measure of this comparison is the interyear correlation. Thus, to derive the continuity of a specific self-image for an adolescent, we can calculate the correlations between his two sortings for the time interval in question. To determine his overall self-image continuity over that interval, we calculate the average interyear correlation of his self-images. The average reflects the second basic dimension of identity formation, *temporal stability* (Hauser, 1971).

This operational definition provides a basis for constructing empirical formulations of variants in identity formation. As Marcia notes in terms of his "identity statuses," different types of identity formation are distinguished in conceptual and clinical discussions of identity development. Each of the types can be translated to a formulation about self-images.

Progressive identity formation is the prototype of normal adolescent development. Throughout childhood and adolescence, identity formation is occurring, with an acceleration of these processes in adolescence. Operationally, we may describe an individual as showing progressive identity formation when the structural integration *and* the temporal stability processes of his self-images are increasing in magnitude, although not necessarily at the same rate.

Identity diffusion applies when there is a clear failure to achieve the synthesis and continuity so essential to identity development. There is a diminishing sense of wholeness and continuity with one's past and with the community. Fragmentation rather than integration seems to be taking place. Empirically, an individual manifests identity diffusion when both processes of his self-image show repeated declines in magnitude over any given period of time. If even one of the processes shows such a decline, we then have an attentuated form of diffusion.

A variant in identity diffusion that superficially resembles progressive identity formation is *identity foreclosure*. At first glance, there seems to be a sense of integration, of "purpose"; and a decrease in subjective confusion about these areas. Yet on closer observation, it becomes apparent that this stability and commitment is but a veneer. The identity configuration is one of impoverished self-definition. The developmental issues have been dodged rather than resolved as the individual has found a rigid characterization of himself, which he now tenaciously holds to. A premature fixing of his self-images has thus taken place. In Erikson's terms, the foreclosed individual does not emerge as "all he could be." Operationally, a person manifests identity foreclosure when either or both the structural integration and temporal stability processes of his self-images show no changes over time.

A fourth identity formation variant is that of *negative identity*. Structurally, this pattern resembles identity foreclosure. There is a premature fixing of self-definition. But in this variant, the focus of the fixed self-characterization is more specific. The individual is committed to those roles and values that have always been represented to him as most undesirable. His sense of identity is based on the repudiated, the scorned. In its premature fixation aspect negative identity does not differ from identity foreclosure. Operationally, this variant appears as identity foreclosure: unchanging self-image processes. To differentiate it from foreclosure requires more information about the content of the fixed images as well as about the individual's own history.

A fifth and final variation in identity formation is that of the *psychosocial moratorium*. This pattern is one of "experimentation" with new roles and images. Rather than settling on any single commitment, any single direction of definition, the growing person "tries on" multiple possibilities. Other terms that are used to define this period are *finding yourself* or *wanderlust*. This pattern represents the opposite pole from identity foreclosure. Instead of being a disruption, an early halt, this configuration is one of openness and change. No irreversible plans or decisions about the future are made. At the same time, the individual is not tending toward or in a type of diffusion. Operationally, an individual is in a period of psychosocial moratorium when the structural integration and temporal stability of his self-images continually fluctuate (increase or decrease) over time.

The operational definitions of the identity variants and their clinical counterparts are summarized in Table 2 and diagrammed in Figure 1. With the meanings of these identity variants in view, we can review results of self-image/identity studies.

Over the past ten years, a variety of findings suggest that this approach meaningfully discriminates among groups of adolescents. In the first study, black and white working-class boys were followed in high school for four years (Hauser, 1971). The results revealed a pattern of identity foreclosure for the black boys and progressive identity formation for the whites. Inspection of yearly clinical interviews conducted with the same subjects gave fuller meaning to these differences. One of the black boys, for example, described recurring work discouragements and visions of a bleak future, one of limited opportunities for him; a picture that would be unaffected by his high school performance. A white subject, in contrast, envisioned many possibilities for "advancement" and change; constantly finding new reasons for optimism about future work and family life (Hauser, 1971).

In a second group of studies, adolescent psychiatric patients were compared with the same age high school students. This time, we found (as predicted) that the patients show clear evidence of identity diffusion as well as more polarized self-images (Hauser 1976b,c; Hauser & Shapiro, 1972). The third investigation incorporating this approach is our current one, following adolescents from diabetic, psychiatric, and high school groups over four years (Hauser, Jacobson, Noam, & Power, 1983). An important feature of this study is that it includes a battery of additional measures (ego development and ego processes), which are *conceptually relevant* to the self-image/identity assessments. Thus, through inclusion of ego development and ego process variables, we now look more closely at the construct validity of the Q-sort as we predict ego development dimensions of the identity variants.

Results from the first two years show striking links between identity variants and ego development levels. Those adolescents at low levels of ego development *also* have diminished self-image integration and temporal stability. Thus, we find a significant link between arrested ego development and identity diffusion. Clinically, this is a meaningful finding. We know from

TABLE 2. Identity Formation Variants

	Definition	
Variant	Clinical	Operational
Progressive identity formation	Continual increase in synthesis of identifications, personal and social continuity. "An evolving configuration of constitutional givens, idio-syncratic libidinal needs . . . consistent roles"	Consistent increase in both structural integration and temporal continuity
Identity diffusion	Continual decline in synthesis of identifications and ego functions; decreasing sense of wholeness and continuity with self and community. Fragmentation	Progressive decline in struc-tural integration and tem-poral stability; more severe form has greater decline of structural integration
Identity foreclosure	Rigidity in self-definition. Lack of any change in synthesis of identifications or other synthetic processes. Char-acterized by premature aborting of identity devel-opment. At first glance resembles "successful" identity formation	Little or no change in tem-poral stability or structural integration. If there is any marked change, it is in the direction of increasing tem-poral stability, in face of unchanging structural integration
Negative identity	Premature self-definition based on *repudiated,* scorned iden-tifications and roles. Com-mitment to what is personally despised	Same as identity foreclosure, of which it is a subtype. Examination of *content* of self-images essential to dis-tinguish it as a subtype of foreclosure
Psychosocial moratorium	"Experimental" state; no firm commitments made. "Trying on" of roles and integrations characterized by flexibility, flux, but *not* disintegration	Fluctuation, swings, in both directions of temporal sta-bility and structural integration

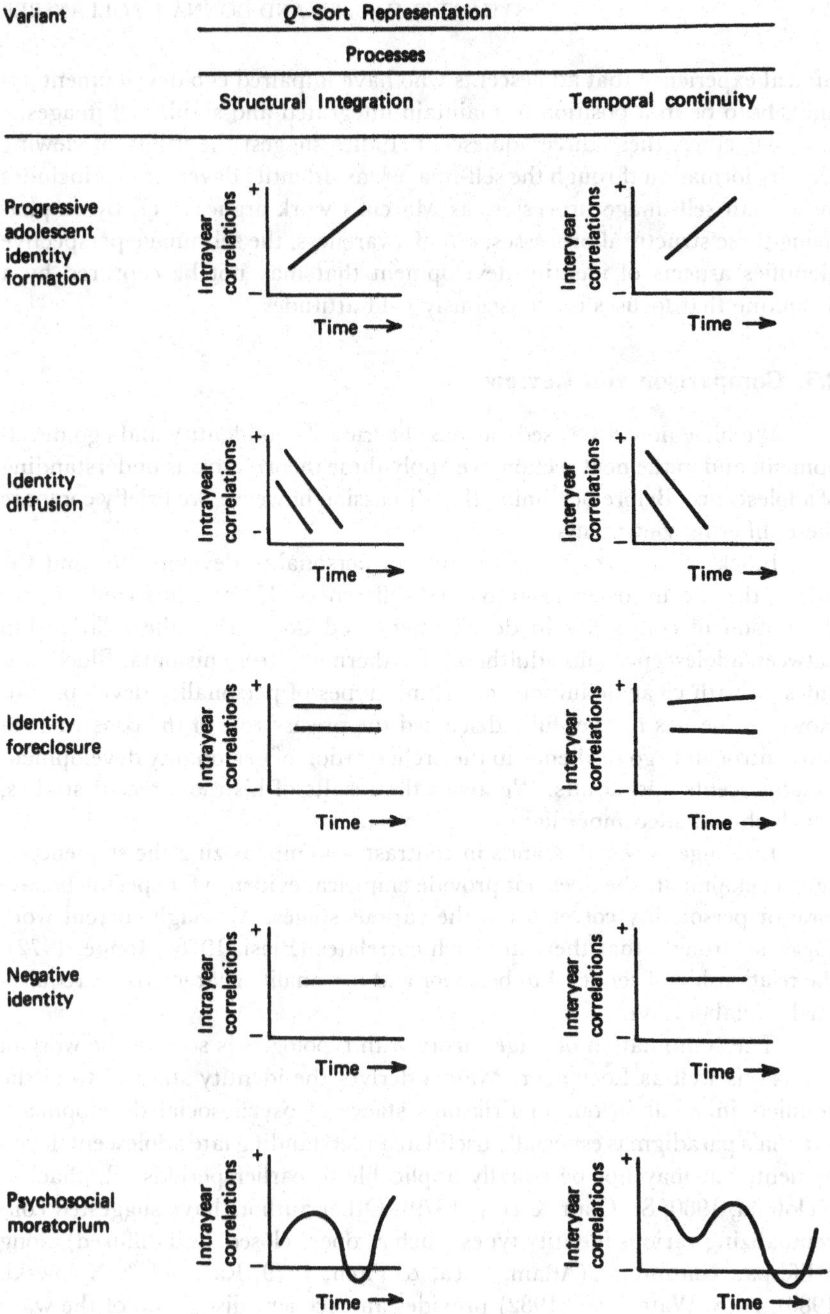

Variant	Q-Sort Representation	
	Processes	
	Structural Integration	Temporal continuity

FIGURE 1. Identity formation variants.

clinical experience that adolescents who have impaired ego development are unlikely to be in a position to maintain integrated and stable self-images.

Together, these three adolescent studies suggest the utility of viewing identity formation through the self-image lens. Identity development includes more than self-image processes, as Marcia's work argues. Yet, by emphasizing these structural processes, out of awareness, the self-image perspective identifies aspects of identity development that may not be captured by a technique that focuses on consciously held attitudes.

2.5. Comparison and Review

We have now discussed various theories of ego identity and ego development, and in the next section, we apply these theories to our understanding of adolescence. Before beginning that discussion however, we briefly compare these differing viewpoints.

Block emphasizes the continuity of personality development and the role of the ego in fostering individual differences. He presents evidence for the notion of continuity in development and documents the relationship between adolescence and adulthood. Furthermore, from his data, Block provides us with clear definitions of distinct types of personality development. However, he has not yet fully discussed the precise role of the constructs of ego control and ego resiliency in the orchestration of personality development in adolescents and adults. We await the results of his most recent studies, which should shed more light on this area.

Loevinger's theory stands in contrast. In emphasizing the sequence of ego development, she does not provide empirical evidence for specific behavioral or personality correlates of the various stages. Although current work suggests strongly that there are such correlates (Blasi, 1976; Hoppe, 1972), the relationship of ego level to behavior and personality characteristics requires further elaboration.

The combination of stage theory with typologies is seen in the work of Marcia as well as Loevinger. Marcia derives the identity statuses from the conflicts inherent in one of Erikson's stages of psychosocial development. Marcia's paradigm is especially useful in understanding late adolescent development, but may not be equally applicable to earlier periods (Raphael & Xelowski, 1980; St. Clair & Day, 1979). Other authors have suggested conceptualizing various identity types (such as open, closed, and diffused) along a lifespan continuum (Adam, Shea, & Fitch, 1979; Raphael & Xelowski, 1980). A. S. Waterman (1982) provides an excellent discussion of the ways in which the identity statuses can be seen as a developmental progression. Thus, the various types can be viewed sequentially as well as individually.

The work by Hauser in self-images also uses longitudinal data to derive an understanding of identity variants. Similar to Loevinger's work, there is an emphasis on the process of self-image formation; but combined with this is a clear depiction of the types of identity that correspond to different self-image processes.

Thus, we see in each of these theories a focus that integrates that underlying process with the individual styles of ego development and ego identity. Each author chooses his or her own point of emphasis: Block emphasizes distinct types, but postulates organizing constructs of ego resiliency and ego control; Loevinger describes more fully the process of ego development, giving much less attention to behavioral or personality correlates. Marcia originally used a typological description, but is now moving to a more detailed discussion of the sequence of development that connects the various identity statuses. Hauser's view is one of sequential growth and change, capturing different sequences of self-images formation within the overall process.

Although each theorist chooses to highlight a particular aspect of adolescent identity development, there are various points of correlation. These are displayed in Table 3, where both types and stages are compared. For example, the stage Loevinger depicts as self-protective is one in which the world is viewed very simply, in terms of "oneself and all others." The individual's sense of identity is not at all complex, and must be somehow insulated from the potential threat of others. Relationships are shallow; the viewpoint egocentric. This description also fits Marcia's identity diffuse status, in which youth are depicted as uncertain of who they are or what they believe, and consequently vigilant and socially withdrawn. One would expect to find very poorly formed, or declining, self-image processes in these individuals, lacking in complexity or stability across time. Such is the case in the individuals studied by Hauser. Finally, specific personality types conceptualized by Block can be seen as having rather diffuse identities, possibly corresponding to Loevinger's self-protective stage during their adolescence: the belated adjusters and anomic extroverts among the males and dominating narcissists and the lonely independents among the females. The difference in Block's groups is that the latter group within each gender goes on to develop greater ego resiliency, whereas the former does not. This type of movement is just one illustration of the need to consider both the ongoing process and the individual styles of ego development and ego identity. As the table indicates, we do not yet have one theory that can adequately provide the conceptual and empirical complexity to encompass the intricacy of ego formation during adolescence, but by maintaining an eclectic perspective, we can gain the necessarily integrated understanding.

TABLE 3. Comparison of Stages in Theories of Ego Development and Identity Formation

Loevinger's ego development	Marcia's identity statuses	Hauser's self-image processes	Block's personality types	
			Male	Female
Presocial				
Symbiotic		Declining complexity of self-image		
Impulsive	Identity diffuse		Unsettled undercontrollers	Hyperfeminine repressives
				Vulnerable undercontrollers
Self-protective		Negative identity	Belated adjusterss	Dominating narcissists
			Anomic extroverts	Lonely independents
Conformist	Foreclosure	Fixed self-images	Vulnerable overcontrollers	Cognitive copers
Conscientious/ conformist	Moratorium	Fluctuating self-image		
Conscientious	Identity achievement	Increasing complexity of self-image	Ego resilients	Female prototypes
Individualistic				
Autonomous				

3. ADOLESCENT DEVELOPMENT THROUGH PHASES OF GROWTH AND FAMILY INTERACTION

3.1. A Series of Growth Tasks

We now apply these developmental perspectives to the period of adolescence. We focus on Erikson and Loevinger, with their rich theoretical foundation for the description of this period. Each of these theories has also been used in empirical investigations of adolescent development. By juxtaposing these theories and the related studies, we can expect to arrive at more detailed understanding of psychosocial development in adolescence.

During adolescence, problems of coherence, continuity, meaningfulness, and self-definition may, and frequently do, take precedence in individual awareness. At times these problems take on overwhelming importance. It is within this area of development that correspondences between childhood expectations and envisioned adulthood are sought, and so often not found. Major reorganizations and integrations may take place, yet many of the changes are gradual, silent, and nontumultuous (Offer, 1969; Offer & Offer, 1975).

Ego processes are closely involved in these changes (J. Block, 1971; Erikson, 1963; A. Freud, 1936; Josselson, 1980). From a psychoanalytic view, ego functions enable an adolescent to sort through and synthesize old childhood identifications, gaining independence and integrating newly autonomous drives, abilities, beliefs, and representations. Ego identity represents the work of the ego in integrating aspects of self into a coherent whole, incorporating those parts that have become individual and autonomous. In terms of Loevinger's perspective, during adolescence there is generally increasing differentiation of the perception of self, the social world, and the relation of one's thought and feelings to others. Erikson and Loevinger cite the important integrative processes: Erikson stressing independence from parental influences and growing autonomy of inner self-representations (Erikson, 1968); and Loevinger highlighting the reorganization of the frame of reference through which an adolescent grasps self and interpersonal experience. The two are by no means incompatible, as they depict both changes in the *unconscious experience* and the adolescent's conscious view of that experience. Whereas the psychoanalytic view (Erikson) stresses internal and external change and the role of ego as mediator, the cognitive-developmental view (Loevinger) emphasizes how various changes may lead to the emergence of different structures of meaning. Drawing on both perspectives we focus on tasks of early adolescence, and move through the middle and late phases, pointing out the challenges and changes of each phase.

3.1.1. Early Adolescence

Before we begin, a general caveat is in order. We do not assume that age and phase of development are consistently coordinated; much work suggests that the correlation between stages and chronological age is variable (Josselson, 1980). Although it makes sense to order our discussion chronologically, the reader must bear in mind that an unusually "precocious" 13-year-old could be at the later phases of ego development.

The years preceding adolescence are ones of relative calm. Oedipal conflicts have for the moment subsided. Sexual impulses are, again for the moment, dormant. This is a time of much practical learning. The school-age child is learning skills that are directly introducing him to the complex technology and thought of his surrounding society. Parents and teachers are seen increasingly as being among the crucial representatives of society. In other words, emergent at this point is a growing awareness of roles and relationships outside the immediate family setting.

With the advent of puberty major shifts occur in psychological as well as physical development. Previously established ideals, rules, defenses, and perceived continuities are now rearranged, or rejected.[9] Besides the internal biological stresses of intensified sexual drives and rapidly advancing bodily changes, numerous pressures of a social nature become crucial. These include issues of new physical and psychological intimacies. And present as well are those issues related to the social setting itself. As adolescents enter the junior high school, academic and peer pressures increase; the comfort and stability of the single classroom and individual teacher vanish. The young person experiences a seemingly urgent pressure as he or she symbolically leaves childhood and enters adolescence (White & Speisman, 1977). Within this turmoil, self-conscious and disquieting questions become more conscious: "What do they think of me? Do they like me? What do I stand for? Where do I belong?" Definition, synthesis, and continuity are the leading themes of this turbulent time.

> The ego of the adolescent is in great need of support, yet paradoxically it has to provide this support out of its own resources. Against newly intensified impulses it has to maintain the old defenses and create new ones; it has to consolidate achievements that have already been reached. The most important of its tasks is the struggle to synthesize all childhood identifications as they become enlarged and enriched by new ones. The successful end result of this struggle will be the formation of a solidified personality, endowed with a subjective feeling of identity that is confirmed and accepted as such by society. (Deutsch, 1944, p. 33)

[9]This general problem of adolescent shifts and "rearrangement" is discussed by Anna Freud in *The Ego and the Mechanisms of Defense* (1936).

Erikson (1963) maintains that the intensifications of impulse (sexual and aggressive) in puberty trigger new phases in adolescent psychosexual and ego development. Others conceputalize ego development as an autonomous process, which does not rely on psychosexual forces to trigger its growth (Josselson, 1980). The precise relationship of biological factors to psychosocial-developmental shifts is intriguing and incompletely understood. Yet we are somewhat clearer with respect to unfolding tasks of early adolescence. The initial tasks include discerning self from other, and the gaining of greater separateness and autonomy from the formerly parent-centered world. To accomplish this, "adolescents have to refight many of the battles of earlier years," even though to do so they must "artificially appoint perfectly well-meaning people to play the role of adversaries" (Erikson, 1963, p. 261).

The first "old battle" to be refought is the struggle for autonomy. In this struggle, an adolescent views parents and other authority figures as adversaries, often delighting in saying "no," as he or she searches to bolster shaky feelings of independence. It is as though these preliminary separations from childhood identifications and representations require some flexing of will, proving through oppositional behavior that one is indeed different from his or her parents and family (Josselson, 1980).

A danger at this early juncture is that such preoccupations with autonomy may lead to an undesired outcome, a prematurely independent youth who has not yet developed inner resources that can replace the support formerly provided by the parent. In these years, peers often serve as transitional sources of support and identification. Secrets, gangs, cliques, and crowds can help guard against the danger of "role confusion"—the dissolution of one's sense of self, role in life, and control of feelings. The search for self-definition is eased by the adolescent's use of peer groups to create a sense of belonging and of being valued.

The focus on peers takes place at a time when the young person is also making increasing differentiations of the world, perceiving greater complexity among groups, lifestyles, and possibilities. The youth's earliest rebellion is often an effort at self-determination; a way to begin making choices in an overwhelming sea of alternatives.

This increasing differentiation of the world is also an aspect of a change in ego development level. As described by Loevinger, the transition from the self-protective level to the conformist level encompasses a growing awareness of one's relationship to others. The self-protective level, often found in preadolescent youth (Loevinger & Wessler, 1970), is characterized by an external organization, in which the world is divided into two classes: oneself and all others. At this stage, one is "somewhat confused . . . having inadequate conception of the complexities of the world" (Loevinger & Wessler, 1970).

The primary preoccupations are with self-protection and staying out of trouble. In social relationships, the youth is wary and manipulating; on guard at all times to control the situation and himself or herself (Loevinger & Wessler, 1970). Usually, persons at this level are not yet cognitively able to take the role of another. Without the understanding of other people's views and reasons for behavior, they must constantly be on guard, protecting their own interests against seemingly capricious whims of others. Social norms and group affiliations are not very meaningful, as the orientation is primarily egocentric.

This view of the world begins to change as the important process of decentering takes place. The early adolescents are just beginning to view themselves as *part* of the world, rather than the *center* of it. The transition is a difficult one. As the youth begins to perceive himself as one small part of a great universe, the need for autonomy grows as a way to derive a sense of self. Simultaneously, peer groups are used in the same quest for self-definition. The changes seen as the individual leaves childhood are accompanied by a changing conception of the world and one's role within it. Classification of people into groups, emphasis on behavior and self-role are, in part, the beginnings of the struggle for identity; but they are also the earliest realization of the most elementary divisions in the world. In other words, the adolescent is changing in response to psychosocial crises, and in his or her basic perceptions of self and others. The youth struggles to become more independent in a world that he sees as more complex.

3.1.2. Middle Adolescence

As adolescence progresses, the youth's perceptions increase even more in complexity and diversity. Young people continue to rely on groups and peers for self-definition. Conformity to peer norms is the dominant concern, and various roles are tried out, in an effort to find those which gain sound approval and feel comfortable.

Young love is part of this process. Erikson (1968) explains that youthful romances contain much conversation because boys and girls clarify their projected self-images through discussions with their loves. Early romances also serve to give the youth a sense of self-worth as a desirable person. Finally, these relationships may fulfill the dependency needs awakened in the adolescent's transition from reliance on parental support toward autonomy.

In the reliance on groups and peers for self-definition, and their use of young loves as a source of meeting transplanted dependency needs, adolescents often manifest the characteristics of the conformist stage of ego development. Peer groups, or "crowds," may enhance tendencies toward holding stereotyped ideas about others. Conventional social norms, stereotypes, clichés, and traditional sex roles are accepted without question. Both relationships and feelings are treated in a similarly superficial, banal manner. Attention is to behavior and surface emotions rather than to inner states.

Although some adolescents may refuse to conform to parental demands, they eagerly comply with peer norms. In this pattern we see a possible connection between ego development, in Loevinger's sense, and ego identity, as defined by Erikson. Whereas Erikson helps us understand the psychosocial and social conflicts of the period, Loevinger deepens our understanding of the adolescent's orientation to the world as these conflicts emerge. Thus, in the middle phase of adolescent-identity acquisition, youth develop autonomy from their parents but are often ready to subordinate their new individuality in the demands of their chosen peer group.

Sometime near the end of high school, or most often at the beginning of college, a shift in focus signaling a preparation for the final phase of adolescence occurs for many youth (Redmore & Loevinger, 1979).[10] Their perceptions become less stereotyped and increase in complexity and diversity as the transition from the conformist to the conscientious level of ego development is made.

Appropriately termed the *self aware* level, this period of transition is marked by heightened awareness of self *separate* from the group, and deepened interest in inner feelings and interpersonal relationships. The world is no longer seen in concrete, absolute terms; groups and individuals are no longer identified through their superficial qualities. Rather, inner states such as motives, feelings, and individual differences are perceived. Interpersonally, the emphasis on belonging and conforming to norms and rules to gain social approval gives way to an emphasis on self-definition and developing one's own standards and ideals. The focus is on individual rules, standards, and ethics; and developing the sense of one's own inner feelings and ideals among the complex choices now perceived. Conceptions of purpose, goals, models, and expectations emerge. Such conceptions become sharper at the next stage, through a growing awareness that one must define one's own path and may need assistance in doing so.

Loevinger states that the modal time for this transition to occur is during the first two years of college (Loevinger & Wessler, 1970). This coincides with what is for many a period of moratorium:

[10]It is important to realize that Loevinger purposely does *not* describe a typical progression through ego stages. Any given individual may be at *any* state of ego development during adolescence; some may remain at the self-protective stage; some at conformist; some may progress to the conscientious level early. The implications of these permutations will be discussed at the end of our descriptions of phases of adolescence.

a period when the young person can dramatize, or at any rate experiment with, patterns of behavior of both juvenile and adult, and yet often find a grandiose alignment with traditional ideals or new ideological trends. (Erikson, 1968, p. 200)

The adolescent's deepening perception of complexity is facilitated through the chance to try on various lifestyles, morals, and ideals. As goals and ideals gain relevance, he must continually exercise self-criticism and evaluation. Moratorium is a mind that vacillates between grandiosity and despair as the youth attempts to tailor higher dreams to match his true capabilities.

Friendships, too, provide a forum for self-definition as one begins to discern differences not only between people but within an individual's fluctuating friendships. Hatred one day and love the next are typical, as the youth becomes better able to recognize and accept positive and negative qualities simultaneously (Josselson, 1980). Inner feelings, too, become more difficult, as the young person gains self-awareness and seeks to answer the question, "Who am I?" This question, at the root of Erikson's "identity crisis," is echoed in the uncertainty, discomfort, and loneliness of youth during this period. As the adolescent becomes more sophisticated in perception, relations, and self-knowledge and struggles through the uncertainty of the new differentiations, the stage is set for the final phase of adolescence and the emergence of identity.

3.1.3. Late Adolescence

The resolution of the identity crisis is the critical task of late adolescence. The result of cumulative growth and change, the identity forged is *not* a stagnant entity; it is a developing commitment, sexual orientation, ideological stance, and a vocational direction. This is the description offered by Marcia (1980), who elaborates phases of identity acquisition. We consider his work in conjunction with Loevinger's in viewing this phase of adolescent development.

Through the attempts at self-definition, and the growing perception of the world's complexity the adolescent is evolving a sense of commitment to himself and to the world. The emergence of this commitment can be seen through Marcia's identity statuses. Recent work suggests that development proceeds from identity diffusion, as the least mature status, to foreclosure and/or moratorium, culminating in identity achievement (Matteson, 1977; Munro & Adams, 1977; Waterman, 1982). In addition, a reported relationship between identity status and ego development indicates that youth who have reached identity achievement tend to be either at the conscientious level of ego development or above it; those at the diffusion status are more often at the self-protective or conformist levels (Adams & Shea, 1979). Hence, research findings suggest that, as the youth attains greater complexity and precision in his or her orientation to the world, advances may also be made in the sense of commitment to self and to others.

These two developmental processes are illustrated in the clinical observations of J. M. Donovan (1975) and Josselson (1973). *Identity diffusion* students are described as generally withdrawn, and feeling out of place in the world (J. M. Donovan, 1975). With regard to politics, religion, and sexual relationships, they are vague, troubled, and confused. Unsure of both preferences and feelings, they seem to be frightened, sad people, with low self-esteem. They are furtive, withdrawn, and guarded with their peers, tending to have the lowest level of intimacy with same-sex friends (cf. also Marcia, 1976a,b; Orlofsky, 1973).

This description is much like that of the women who comprise Josselson's (1977b) group. These women also expressed low self-esteem, feelings of alienation, and inability to commit themselves to anyone or anything. Josselson notes their fantasy that they could become *anything* and their fear that to commit was to give up that possibility. Feelings were not used for meaningful contact with others but rather were the basis of the everchanging, romanticized, self-centered fantasy life, which was at the core of these females' identity.

It is understandable that within the identity diffusion status we find the greatest number of individuals at low levels of ego development. In fact, Adams and Shea (1979) report the majority of identity diffusion adolescents are at the preconformist level. The furtiveness, withdrawal, and alienation, may in part be related to an inability to see the world from other's viewpoints, or to understand why others feel and act as they do. The wary, manipulative quality of the lower ego levels corresponds to the behaviors described by Josselson and Donovan. These behaviors and attitudes may reflect a basic orientation to the world; in which commitment to another would mean dissolution of self.

The dominant theme for both young men and women who fit the *identity foreclosure* status is close affiliation with parents and family. Adolescents in this group status have internalized parental standards and ideas, choosing jobs, lifestyles, and religious beliefs that directly parallel those of their parents. They have never reworked what was offered by their parents, possibly because to do so would require rejection of an abundant supply of warmth, affection, and love. J. M. Donovan (1975) suggests that sensing oneself in disagreement with a parent causes these youth to feel guilty and alone. Similarly, Josselson (1980) notes that women in the foreclosure status have not been able to trust others enough to form the peer relationship necessary to replace parental identification. The sole source of gratification and self-esteem therefore remains anchored in the parent-youth relationship. Both men and women fail to express strong feelings for others and maintain relations primarily for the purpose of getting what they want. Although purposive, talkative, constructive, and task oriented, the men seem not to be independent, curious, or aware of their own thoughts and feelings. The women are superficial, conforming, and sweet. Women and men alike have a self-assured, goal-oriented quality as sober, responsible young adults. They strive to create the type of home and society that their parents designed.

The conformist level of ego development is the modal one for foreclosure individuals. Recalling the previous description of this level, one can easily see why this would be so: just as the foreclosed status is characterized by unquestioned adherence to parental standards, the conformist level is one of unquestioned obedience to rules. The conformist's emphasis is on actions rather than feelings, and interpersonal reciprocity is based more on a "golden rule" mentality than on true mutual trust. Similar to this is the foreclosed individual's difficulty establishing trusting relations, and the self-centered, stereotypic, structural quality of their interactions with others. Concern with reputation and status, appearance, and adherence to structural goals are also preoccupations of both identity foreclosure and conformist individuals. Not surprisingly, 63% of the foreclosed individuals in the Adams and Shea (1979) studies were at the conformist level of ego development.

The *moratorium* individual is in the midst of crisis and conflict. Indecision manifests itself everywhere: occupational choice, sexual standards, and personal and political ideology are all in flux. Struggles may stem from attempts to become free of parents: women especially tend alternately to declare independence and then return, fearful that autonomy will be coupled with complete rejection (Josselson, 1975). Interpersonal relations are intense but ambivalent. These individuals are described as competitive, counterdependent, and hostile, but also visible, engaging, and emotionally responsive (J. M. Donovan, 1975). They are ever engaged in struggle for control of their social group—and establishing a central place for themselves within it. Unable to maintain close relationships, they tend to look to others to give them the approval and assurance they cannot give themselves, but then are threatened by any loss of distance from the other (Josselson, 1973). Feelings are experienced and expressed with depth and intensity, and in general seem to be the epitome of what adolescence is traditionally considered to be:

> They have the acuity of vision, the responsiveness to social problems, and the psychological closeness to the great philosophical issues of life that have so often been sentimentalized by observers of youth. (Josselson, 1973, p. 34)

In terms of ego development, we see a slight shift upward in the moratorium as compared to foreclosures. The self-awareness, the greater focus on feelings in relations, and the overall vivid intensity that marks the youth in moratorium are characteristic of those at the transition from conformist to conscientious levels, and of the conscientious level itself. The increased perception of complexity, multiple choices, and contingencies that accompany the shift from conformist to conscientious stages could easily be seen as part of the confusion and turmoil of the moratorium adolescents. Deepening interest in interpersonal relations and concentration on life purpose, goals, and meaning are also characteristic of growth in ego development and identity acquisition.

The identity achievement status is one that represents the consolidation of the identity components. A sense of autonomy, an examined vocational direction, and a sexual orientation are affirmed. Crisis has been encountered and largely resolved, and commitment to oneself and the world is emerging. Unique to this status is the ability to develop an autonomous sense of self-esteem and identity. This is illustrated by the contrast between the foreclosure pattern, in which parental values are accepted totally, and the achievement pattern, in which some aspects are accepted and others redefined or rejected. Nor is the redefinition of ideals and standards accompanied by the guilt of the moratorium pattern. Rather, youth of the identity achieved status are able to turn to their own capacities, talents, and beliefs as sources of support and self-esteem; they need not rely wholly on either parents or peers. Their ability to appraise others realistically and to tolerate ambivalent feelings of both love and dislike reflect their increased facility on confronting complexity. Having struggled with numerous experiences that attested to their ability to be on their own and survive, they are committed to having their own lives. Interpersonal relations are rich and peers are used to facilitate independence: for example, females do not choose parental substitutes to "care for" them in replacement of parents, but rather choose men who will "care about" them to replace the self-esteem lost as distance from the parental source increase. Relationships and work are used to replace the old object ties and unfulfilled wishes (Josselson, 1973). From these sources, people in the identity achieved status derive satisfaction and awareness of their own capacities. Although they may maintain lofty goals and dreams, they are able to temper their aspirations with realistic assessments of their needs and abilities.

The data of Adams and Shea (1979) suggest that identity achievement is associated with higher stages of ego development. As was the case with all other statuses, there were people functioning at various ego stages within the achievement category. It is noteworthy though that *none* of the individuals were at the self-protective stage; the lowest were at the transition to the conformist stage.

One interpretation is that the ability to integrate seemingly disparate feelings and perceptions, the growth of reliance on inner standards, and the deepening importance of relationships that are part of achieving identity are also indicative of higher levels of ego development; and would be expected to evolve somewhat simultaneously. It would seem difficult, therefore, for an individual to confront the crisis and commitment of identity achievement without having gained a more complex awareness of self and others.

In our description of the phases of adolescence, we have integrated psychoanalytic and cognitive-developmental viewpoints. Although different in focus, the two theories complement each other and yield a fuller understanding of adolescent development.

3.1.4. Gender Differences

Thus far, we have not separated males from females in discussing identity formation and ego development. Various reports suggest, however, that there are differences for men and women in the context, and possibly in the process, of identity formation (J. F. Block, 1973; Josselson, 1973; Marcia, 1980; A. S. Waterman, 1982). We have already mentioned the possibility that ego stability is more crucial for females than for males, possibly making the choice of a "foreclosed identity" relatively *more* adaptive for women. One suggested reason for this comes from the work of the Blocks (J. Block & Block, 1977), who have found that socialization processes tend to narrow sex-role definitions and behavioral options for women but widen them for men. On this basis alone, it is predictable that the psychosocial process of identity acquisition should more easily lead to further self-definition, restlessness, and rebellion for men (characteristic of moratorium), and to submission, conservatism, and conformity (characteristic of foreclosure) among women.

With respect to ego development, such socialization would seem to be inhibiting for women. Block (1973) reports two studies that examined Loevinger's stages of ego development and sex-role socialization. Together, the studies indicate that higher levels of ego functioning are consistent with blending both masculine (agentic) and feminine (communal) traits.

These findings are also pertinent to the topic of identity formation. The blend of feminine and masculine concerns occurs sequentially in Erikson's original model, through resolution of the identity and intimacy stage crises. Some authors suggest however, that for women, the concerns of the intimacy stage punctuate identity formation, resulting in a blend of the two stages (Josselson, 1973; Josselson, Greenberger, & McConochie, 1977; Marcia, 1980). Hodgson and Fischer (1979) conclude that the male adolescent tests who he is through issues of competence and knowledge. He seeks to develop a sense of self as a contributing member of the adult world. For the female, however, identity issues are ones of relating, as she seeks to determine a sense of self-finding ways of getting along with others that will satisfy her needs and theirs (p. 47).

Two sources of these differences are possible. One problem may be in Erikson's original conceptualization. His model was developed to be applicable to men and was then extended to women. It may be that the insights he has provided into feminine identity require greater examination through systematic research and theory development. He did *not* postulate different developmental processes for males and females (A. S. Waterman, 1982) but did acknowledge that much of female identity is determined by intimate relationships (Erikson, 1968). Erikson suggests that anatomy dictates that a great part of female identity is determined by her use of interpersonal skill to find a mate. He further suggests that a female remains in moratorium until she accepts a man: "A true moratorium must have a term and conclusion: Womanhood arrives when attractiveness and experience have succeeded in selecting what is to be admitted to the inner space for keeps" (Erikson, 1968, p. 283). It may be, then, that identity formation takes longer for women than for men, and the precise process should be examined more closely. This is the suggestion of Marcia (1980), who contends that a greater elaboration of Erikson's theory may be necessary in order to fully understand female identity formation.

The second possible source of noted differences is cultural emphasis on affiliation for females, and occupation/ideology for males. Many researchers (E. Donovan & Adelson, 1966; Josselson, 1973) have interpreted females' focus in the identity process to be on interpersonal relations, with work concerns also framed in these terms: "I want to use the law to help people," as opposed to "I want to be a lawyer in a corporation law firm" (Josselson, 1973, p. 49). To the extent that this emphasis is culturally determined, we may be seeing differences in the focal concerns of identity formation that reflect cultural adaptation, rather than separate patterns of identity development. As the areas of occupation and ideology, and the importance of independence and self-reliance become more culturally valued for women, fewer differences in the focal issues and in the process of identity acquisition may begin to be seen.

The questions to be answered therefore remain: Do men and women engage in a different *process* of identity formation, with intimacy preceding or merged with identity concerns for women? Such is the conclusion reached by authors who have reported the emphasis on interpersonal relations among adolescent females (Constantinople, 1969; Josselson, 1973; Josselson, Greenberger, & McConachie, 1977b; Marcia, 1980). Or, do men and women engage in the same process, with different focal concerns and different meanings attached to the forms of identity formation (i.e., the identity statuses)? This latter question is suggested by Waterman (1982) who states:

> It is clear that males and females are more similar than different in their use of developmental processes, but their is reason to believe that the identity statuses may nevertheless have different psychological implications for the sexes. (p. 351)

This interpretation integrates both the possibility that the foreclosure status is more adaptive for women, and the conclusion of no sex differences in the many areas of identity formation defined by Marcia (summarized by A. S. Waterman, 1982).

In conclusion, we recognize that ego development is optimized by an integration of what have traditionally been separate "masculine" and "feminine" emphasis. It is also necessary to recognize that the *process* of identity formation, arriving eventually at an integration of identity with intimacy issues for both men and women, has not been conclusively delineated for both sexes. It may be that the process evolves very differently for the two genders. Or, it is possible that current cultural forces act to create different focal issues for men and women, leading to different forms of adaptive resolution of the "identity crisis." Additional empirical research is required to provide a more detailed map of identity formation for both males and females.

3.2. Family Contexts of Adolescent Ego Development

Until now, we have concentrated on *individual* adolescents: their identity formation, shifts in ego processes, self-images, and continuities over time. In our brief discussions of racial and gender differences, we approached interpersonal and societal aspects of adolescent development. Only an exclusively biological model of adolescent development can afford to bypass these social dimensions. To gain a full understanding of psychosocial development in adolescence, one must take account of the immediate social context (family and peer groups) as well as the larger social settings in which these groups are embedded (school, work, ethnic, racial, and social class). In this section we review what is known of *one* set of social influences: the family.

In terms of development, the family plays a special role in the years preceding adolescence. This is the setting for early dependency ties, later complex identifications, sibling interactions, first skill learning, and initial separations and losses. Despite general agreement about the family's significance, and a sprawling family-process literature, there are few specific measures and theories that directly capture the ways family interactions relate to ego development in latency and adolescence.

A number of family processes have been the focus of considerable interest to psychologists and psychiatrists over the past twenty years. Among these are the double-bind (Bateson, 1972; Bateson, Jackson, Haley, & Weakland, 1956); enmeshment (Minuchin, 1974; Minuchin, Rosman, & Baker, 1978); pseudomutuality (Wynne, Ryckoff, Day, & Hirsch, 1958); and family schism and skew (Lidz, Cornelison, Fleck, & Terry, 1957).

The body of theory that has emerged from these efforts is impressive, yet its relevance for the study of ego development is seriously limited. There are at least two related reasons for this: (a) these family theories have been derived largely to explain the etiology of schizophrenia, and certain other forms of psychopathology, "as the product of specific patterns of family interaction" (Broderick & Pulliam-Krager, 1979); and (b) the data base for the family models consists primarily of clinical observations of diagnosed psychiatric patients (usually schizophrenic). A few studies have looked at families of delinquent or antisocial children, comparing them to normal (nondelinquent) and sometimes schizophrenic families (e.g. Hetherington, Stouwie, & Ridberg, 1971); Stabenau, Tupin, Werner & Pollin, 1965). Besides narrowing the range of children who can be meaningfully compared in terms of their family patterns, these more clinical studies do not lead toward operationalizing many of the theoretically significant family interaction concepts.

A second, large set of family studies are more oriented toward empirical analyses, as they are based on direct observation and self-report data (Jacob, 1975; Riskin & Faunce, 1975). Among these many investigations are several that analyze interaction patterns of families and their adolescent children (e.g., Alexander, 1973a,b; Ferreira & Winter, 1968a,b; G. H. Frank, 1966; Goldstein, Rodnick, Jones, McPherson, & West, 1978; Haley, 1967a,b; Hauser et al., 1981; Hetherington et al., 1971; Jacob, 1974, 1975; Lewis, Beavers, Gossett, & Phillips, 1976; Mead & Campbell, 1972). Five projects are especially relevant to questions regarding the interplay between family milieu and adolescent development. In the remainder of this section, we discuss these studies so as to focus and elaborate key issues in the complex area.

3.2.1. Family Interaction of Adolescent Delinquents

Hetherington and co-workers (1971) compared the family interaction patterns found in families with delinquent and nondelinquent adolescents, using direct observations and questionnaires. The delinquent sample was subdivided into three types: unsocialized psychopathic, neurotic-disturbed, and socialized-subcultural. These delinquent differences are likely to correspond to ego development distinctions (S. Frank & Quinlan, 1976). Although such developmental implications were not taken up by Hetherington, they are of interest here, especially since the family patterns varied across the three subtypes. The family interactions, generated through a special discussion procedure, were analyzed in two major ways. A first set of analyses used behavioral measures assumed to indicate dominance and conflict. The second set was based on ratings of parental attitudes toward discipline. The entire family discussion was scored for such dimensions as warmth, hostility, positiveness of expectations for the child, and anxious emotional involvement.

Although behavioral measures differentiated the delinquent families, the strongest findings were in terms of parent attitudes. Besides showing differences among the delinquent subtypes, the attitude studies demonstrated significant contrasts between parents of delinquents and those of nondelinquents. On many measures, the normal (nondelinquent) parents differed from the parents of all delinquent subgroups. The normal parents were less hostile and "power assertive," had more positive expectations for their children, were less anxiously involved, and were warmer than the parents of delinquents. Results point toward a strong relationship between parent views and adolescent "outcomes." They do not clarify causal links or sequential questions, such as whether the parent attitudes *preceded* the delinquent behaviors or were reactive to them.

3.2.2. Defensive and Supportive Communication in Families of Adolescents

Alexander (1973a,b) also studied families of delinquent (runaway) and normal adolescents. Family discussions generated through a revealed differences procedure (Strodtbeck, 1958) were scored along the lines of "defensive" and "supportive" communications. Defensive communications consist of "verbal and nonverbal behaviors that are threatening or punishing to others and reciprocally invite and produce defensive behaviors in return." These behaviors lead to, "a climate where solutions to problems are seriously curtailed or impossible to achieve" (Alexander, 1973a, p. 613). Characteristic defensive behaviors include actions that are evaluative, controlling, indifferent, or superior. In contrast, supportive communications (genuine information seeking and giving empathic understanding, spontaneous problem solving, and equality) tend to produce lowered anxiety and clearer communication in others.

In two different studies (1973a,b), Alexander found defensive and supportive patterns that varied with families of different types of adolescents. The first findings indicated a relationship between preadolescent (9–11-year-old) aggressive sons and mothers' defensiveness. In a later study, families of runaways showed consistently higher rates of defensiveness than normal families in their discussions. The family patterns were more complex for

supportive communications. Here the "runaway" adolescents were consistently less supportive to their parents than the normal adolescents, although
the parents of the two types of families did not differ in terms of supportiveness. A consistent finding, in both studies, was that parents and children
in normal families were strongly reciprocal in supportiveness but *not* in defensiveness. Again, we see correlations between family processes and adolescent
behaviors. Yet the important questions surrounding determinants are still
unresolved. As we will argue, only through approaches that follow children
and their families over time, can we identify causal pathways.

3.2.3. The UCLA Family Project: Familial Processes and Adolescents At-Risk

The overall research objective of the UCLA Project is, "an assessment
of the patterns of intrafamilial personalities, behaviors, and interactions which
might contribute to the development of psychopathology" (McPherson, 1970).
More specifically, the goal is one of studying intrafamilial relationships in
adolescents who vary in their degree of risk for schizophrenia and subsequent
schizophrenic-like conditions (Goldstein *et al.*, 1978). There are a number of
ways in which this research is relevant to our focus on adolescent ego development: its sample consists of adolescents and their families selected from a
clinic population; many of the methods are based on direct observation of
families; and the adolescents are classified along dimensions other than psychiatric, diagnostic, or delinquency categories. Finally, this group uses a
follow-up procedure for the individual adolescents, thus permitting observations of subsequent development.

Adolescents and their families were studied intensively at the time of
their recruitment from the outpatient clinic. The adolescent patients are also
seen again in two follow-up phases (at five and ten years). The UCLA studies
do not assume or argue that the studied familial processes are determinants
of adolescent psychopathology or later young adult outcome. In fact, several
of the project papers discuss the possibility that the "parental style" findings
are *reactive* to the child rather than indicative of long-standing modes within
the family that have in some way shaped the adolescent (Alkire, Goldstein,
Rodnik, & Judd, 1971; McPherson *et al.*, 1973, 1980). Since this is a longitudinal issue, the cross-sectional family data of the UCLA analyses are similar
to those of Alexander and Hetherington and consequently cannot clarify
which family interactions are antecedent and which are reactive ones. Although
their design includes 5- and 10-year follow-ups of the index adolescent cases
into early adulthood to trace their subsequent histories of adaptation, the
families are not observed at these later points.

The UCLA Project uses several family discussion procedures and analyses keyed to various family dimensions. Their earliest analyses looked at

styles of social influence used by patients and adolescents in resolving family conflicts (Alkire et al., 1971; Goldstein et al., 1978). These first studies were based on individual interviews with parents and their children, using structured interviews and simulated interaction sessions. Influence strategy coding indicated that two aspects of parental behavior were significantly associated with the form of adolescent psychopathology: (a) the overtness of assertiveness used in exerting parental influence; and (b) the focus of power through one or the other parent (Alkire et al., 1971).

Other analyses of these family discussion observations have looked at acknowledgment, or responsiveness, levels using a modification of the Mishler-Waxler acknowledgment scale (McPherson et al, 1973; Mishler & Waxler, 1968). Once again, the findings indicated a systematic relationship between parental style of communication and adolescent problem expression. For instance, these adolescents with more active behavior problems (aggressive, antisocial, and active family turmoil) experienced more direct acknowledgment of their remarks than the more passive groups (passive, negative, and withdrawn). The withdrawn adolescents were continually questioned and interrupted in midspeech, but their parents rarely failed to acknowledge their statements at some level. The most extreme nonacknowledgment occurred for the passive-negative adolescent, whose parents continually questioned him *and* then failed to acknowledge his responses.

In another study, Jones and his associates (1977) applied acknowledgment scales to observation of *low-* and *high-risk families* defined through independent TAT analyses of communication deviance (inability of the parent or parents to establish and maintain a shared focus of attention during interpersonal transactions; Wynne & Singer, 1963). The low-risk families differed from the other two groups (intermediate and high-risk) by their marked frequency of acknowledging and "extending" the remarks of the previous speaker as well as their low frequency of total disregard (Goldstein et al., 1978). Recent contributions from this project concentrate on familial interactive patterns predictive of schizophrenia-like outcomes in a 5-year follow-up study of this initial adolescent sample (Doane, West, Goldstein, Rodnick, & Jones, 1981; Goldstein, 1981). A second focus in current papers is on interactive and outcome correlates of communication deviance.

The overall thrust of these papers is to determine family patterns predictive of the later development of psychiatrically disabled adolescents. The interest in prediction of course signals a concern with "causal" patterns: Do certain family processes underlie later severe developmental impairment (schizophrenia)? In the final two investigations, we continue to examine the important question of family "impact." In addition, the following studies return to the topic of "normal" and accelerated psychosocial development, in contrast to the more intensive focus here on a psychiatric sample.

3.2.4. Family Process Predictors of Adolescent Identity

In a series of studies with high school seniors, Grotevant and Cooper (Cooper, Grotevant, Moore, & Condon, 1983; Grotevant & Cooper, 1982) follow the influence of family process variables on selected aspects of adolescent development (identity and role-taking skills). The conceptual model that they use focuses on "individuation," a property of the family system observable in family communication. Drawing from theoretical, research, and clinical work in family sociology, psychiatry, therapy, and developmental psychology, Grotevant and Cooper (1982) propose that, "the development of identity and role-taking skills in adolescence is most facilitated by an individuated family system" (p. 54). More specifically, the concept of individuation refers to three specific interaction qualities: self-assertion, validation, and permeability. Many of their studies, then, examine the interplay between this profile of interaction qualities and individual development.

Similar to the preceding projects, *both* the individual adolescents *and* whole families are observed through individual measures (e.g., identity interview) and family tasks ("Plan Something Together"). This project, with its explicit measures of adolescent identity development, is thus most pertinent to our considerations of familial contributions to adolescent psychosocial maturation. First findings have been described in a variety of publications, and are summarized in their most recent report (Grotevant & Cooper, 1982).

Identity development was assessed through a revision and extension of the Marcia interview (Grotevant et al., 1982), which we described earlier. The major revision/expansion was through the addition of probes regarding friendships, dating, and sex roles. Each of the adolescent subjects was interviewed by a trained interviewer. The tape of the interview was then rated for exploration, commitment, and "identity status." Analyses of the relations between the identity measures and family process are based on the exploration and commitment scores for individual development; and on conceptual as well as factor analytic family scores. In summarizing the complex results, we highlight the most striking relationships between adolescent identity and family patterns' scores.

Adolescents who were high in *identity exploration* expressed both separateness and permeability in the family discussion. Their fathers expressed mutuality in family interactions, especially through initiating compromises. These adolescents also expressed greater permeability to their fathers than did the low-exploring adolescents, suggesting that this pattern of "fathers supported communication and reciprocated permeability from the adolescent creates a context conducive to exploration" (Grotevant & Cooper, 1982, p. 90).

Mothers who showed more permeability behaviors were more strongly correlated with low-exploration adolescent scores. The causal links are unclear here. Were mothers, who were oversolicitous, preventing their adolescents from developing autonomy? Or could this be explained as the *result* of mothers sensing greater needs or difficulties in their adolescent's (identity exploration) development?

Adolescents high in identity commitment (certain of identity decisions) made relevant comments, and showed permeability, and separateness in the family discussion. Their parents had a low tolerance for expression and resolution of differences, such as initiating compromise or indications of separateness. A possible interpretation of this pattern is that, "the climate of families which avoid the expression of differences inhibits adolescent exploration and stimulates the development of early, possibly defensive, commitment" (Grotevant & Cooper, 1982, p. 91).

Two clear messages emerge from these many findings linking family behaviors and identity development patterns. First, they illustrate and emphasize the importance of the family context in adolescence. Second, they dramatically point out that formulations about family processes and adolescent development are likely most accurate, and testable, when phrased in discrete, "molecular" terms rather than in global "molar" ways. The findings we summarized concern mother-adolescent or father-adolescent behaviors. Others, not presented here, involve mother-father interactions (Cooper *et al.*, 1983). Such complexity is glossed over by any general descriptions regarding the family's "individuation" and its contribution to adolescent identity development. The next, and final, family-adolescent project is, by design, responsive to this second point.

3.2.5. Family Processes and Adolescent Ego Development

Our own study (Hauser *et al.*, 1982, 1983, 1984) is an ongoing one. It has parallels with Grotevant and Cooper's work through a shared interest in adolescent development, and in our use of conceptually derived family variables. In our 4-year longitudinal project, we follow ego development and ego identity (through self-images) of adolescent patients and high school students, together with interaction patterns within their families. The patients are drawn from psychiatric and diabetic groups. The design of the study allows us to distinguish the trajectory of the adolescent's development. Our family analyses explore such questions as: What types of family interactions are linked to varying levels of adolescent ego development? How does the level of parent and adolescent ego development affect family interaction? We investigate the possible role of family processes as determinants of illness outcomes and psychosocial development. But it is also possible that the effects proceed in both directions. Our longitudinal analyses will allow us to investigate the traditional assumption that family interactions shape or determine ego and identity development; as well as the intriguing reactive possibility that the level and rate of adolescent development determine shifts in family interactions. Through repeated observations over time, family interactions that are longstanding and precede shifts in the adolescent's development can be distinguished from those family patterns that emerge subsequent to changes in the adolescent. Our system for analyzing relevant family interactions is the Constraining and Enabling Coding System (CECS) (Hauser *et al.*, 1984). The CECS operationalizes variables derived from a revision and expansion of Stierlin's (1974) theory of how parental interactions affect adolescent maturation. The central point in Stierlin's analysis is that, within disturbed families, members make numerous attempts to interfere with more autonomous and differentiated functions of one another. These impediments to more independent perception and cognition occur through "binding" or constraining interactions, in which certain family members distract, confound, or actively resist differentiation of other family members. We have built on this theory in several ways: first, we conceputalize various types of binding or constraining; second, we study *both* the adolescent *and* his or her parent; and perhaps most significantly, we look at "facilitating" interactions in which family members encourage and support expression of more independent thoughts and perceptions. Our predictions were that adolescents at lower levels of development will be in families where they *and* their parents relate through multiple constraining processes; in contrast to adolescents at more advanced stages. In this more mature group, we expect to see many reflections of enabling processes, such as explaining, curiosity, and acceptance.

We generate family-interaction observations through a revealed-differences procedure, in which the family is asked to defend individual positions about moral dilemmas and to attempt to reach a consensus that would represent the entire family. The transcripts of these taped discussions serve as the data base for our interaction analyses, where trained coders apply the CECS scales to family transcripts.

Preliminary analyses of 14 families from the first year of the study reveal that, within families of arrested ego development adolescents, the parents *and* the adolescent express more constraining speeches (distracting, indifference, and devaluing). On the other hand, within families of unimpeded adolescents both parents and adolescents engage in more enabling interactions, such as problem-solving, explaining, and acceptance.

In addition to constraining and enabling, we rate adolescent *changes* within the family discussions. Here we expect to find that adolescents whose development is unimpaired will show more signs of *progression*, where successive positions are more complex and contribute more to the ongoing discussion. Our reasoning also extends to the impact of the adolescent on his or her family, as we anticipate seeing strong relationships between progressing adolescents and enabling parental speeches. The adolescent change will thus evoke more enabling from his or her parents. In contrast, adolescents who have low ego development scores, or show identity diffusion and foreclosure, will express more regressive (decreased complexity and interfering with discussion) or foreclosed (redundant) speeches. This type of change will, in turn, lead to more discouragement, expressed through diminished enabling and/or constraining speeches from the parents. These speculations will soon be tested. And, similar to the Grotevant and Cooper work, we will begin to investigate other "mapping" of family dimensions onto developmental ones (ego identity and self-image change).

These projects, along with the more clinically oriented *No Single Thread* (Lewis *et al.*, 1976) underline the importance and complexity of family context in adolescent development. There are many possibly relevant family dimensions, families have multiple subsystems that must be taken into account (e.g., father-mother, adolescent-mother, and siblings), and direction of influence cannot be taken for granted. Each of these studies will likely clarify aspects of these issues and thereby contribute to conceptual and methodological advances in our grasp of familial contexts of adolescent development.

4. CONCLUDING COMMENTS

We have viewed major theories of adolescent personality development from several angles: conceptual frameworks, scientific support, and clinical relevance. As we moved through theoretical and empirical complexities, we paused to highlight and reflect on points of convergence and unresolved questions. As an approach to drawing together this vast expanse of detail, we briefly review some of the most salient topics touched on in the preceding sections.

4.1. Development and Personality

We began by considering Erikson, whose name has been so closely associated with the culturally popular "identity" terms. Eminently readable and penetrating, Erikson has contributed in both popular and technical directions to our views of adolescent development. Starting from a psychoanalytic assumption of the importance of psychosexual determinants, Erikson reaches to consider peers, family, community, and social institutions within a broad psychosocial framework. We have pointed out that the identity concept is a many-faceted one, suggesting a number of variations in adolescent development. Marcia discusses these identity types as "identity statuses"—achieved identity, identity foreclosure, diffusion, or moratorium; Hauser conceptualizes the types as identity variants, differing in terms of underlying dimensions of self-image continuity and integration.

The theme of personality patterns embedded in a developmental framework is echoed in Loevinger's work. Here the view is most explicit, as there is a character type that corresponds to each stage of development. Finally, the Blocks delineate an array of personality types within adolescence; each with varied antecedents as well as later sequelae in their adult years. These conceptual and empirical contributions favor a position that views adolescent personality and developmental dimensions intertwined in complex and important ways.

4.2. Social Contexts of Adolescence

Throughout the chapter, we reiterated the obvious yet important point that adolescent development occurs within complex social surroundings. To be sure, various biological determinants are significant. But these forces, be they aggressive or sexual, cannot account for all the variation. Inner drives are shaped by diverse influences, which include family processes, cultural values, and friendships. We could not cover the entire topic of society's influence on adolescent change and chose instead to look most carefully at what is known of the social agent, the family. Through such concepts as family "individuation," and familial "constraining and enabling" we are at the beginnings of identifying theoretically relevant familial processes.

Other societal influences are illustrated through the gender differences found by various authors. A noteworthy, but largely unanswered, question is whether the paths of ego and identity formation differ for men and women. Loevinger addresses this point and concludes that, although females may reach later stages somewhat earlier than males, by adulthood no significant differences in level of ego development are found. By juxtaposing the findings of Block and Marcia, we discover an interesting parallel. The suggestion that the foreclosed identity status is more comfortable for women is analogous to Block's finding that overcontrol was a common aspect of many of their adolescent female personality types. The underlying commonality of both findings may be societal influence, which tends to hold control and stability as important aspects of being female. Future work must address not only the possibility of existing differences between men and women in developing a sense of identity but must also evaluate the impact of a changing society on these processes. Many questions remain unanswered in this realm. In addition to the sheer complexity of the issues at this interface between personality development and social context, separation of the relevant disciplines further impedes progress here. Developmental psychologists often study adolescence in relative isolation from sociologists and anthropologists, who are looking at this same phase of growth. Theories and studies attempting to integrate individual (biological and cognitive) and social (family process and social class) variables are relatively sparse—consequently limiting our understanding of the contextual aspects of adolescent development.

4.3. Deviance and Adolescence

This topic received the least attention within the chapter. Pertinent here are questions that address juvenile delinquency, psychopathology, and medical illness within adolescent development. The general issue covered by such questions concerns both the impact of developing *on* these behaviors *and* the influence of the deviant medical (diabetic) and psychiatric aspects of adolescence (e.g., Goldstein *et al.*, 1978; Hauser *et al.*, 1982, 1983, 1984; Weiner, 1980). Given the current high level of interest in "health psychology" (e.g. Stone *et al.*, 1979), we expect that this topic will receive increasing attention over the next years. In the area of ego development, for example, there is already an indication that its relationship to forms of psychopathology has become a lively topic for several investigators (Gold, 1980; S. Frank & Quinlan, 1976; Noam *et al.*, 1982; Vincent & Vincent, 1979; Waugh & McCaulley, 1981).

It is apparent that adolescent ego identity and development is a sphere of inquiry that has attracted a broad array of clinicians and social scientists. We envision the emergence of many intriguing tasks in future investigations— ones that range from integrative theory building to greater understanding of development aspects of adolescent illness. Our hope is that this review has stimulated readers to consider these and other issues as they appear in clinical and in research settings.

5. REFERENCES

Adams, G. R., & Shea, J. A. The relationship between identity status, locus of control and ego development. *Journal of Youth and Adolescence*, 1979, *8(1)*, 81–89.

Adams, G. R., Shea, J., & Fitch, S. A. Toward the development of an objective assessment of ego identity status. *Journal of Youth and Adolescence*, 1979, *8(2)*, 223–237.

Alexander, J. F. Defensive and supportive communication in normal and deviant families. *Journal of Consulting and Clinical Psychology*, 1973, *40*, 223–231. (a)

Alexander, J. F. Defensive and support communications in family systems. *Journal of Marriage and the Family*, 1973, *35*, 613–617. (b)

Alkire, A., Goldstein, M., Rodnick, E., & Judd, L. Social influence and counterinfluence within families of four types of disturbed adolescents. *Journal of Abnormal Psychology*, 1971, *77*, 32–41.

Ansbacher, H. L., & Rowena, R. (Eds.). *The individual psychology of Alfred Adler: A systematic presentation in selections from his writings.* New York: Basic Books, 1956.

Bateson, G. *Steps to an ecology of mind.* New York: Ballantine, 1972.

Bateson, G., Jackson, D. D., Haley, J., & Weakland, J. H. Toward a theory of schizophrenia. *Behavioral Science*, 1956, *1*, 251–264.

Bernfeld, S. Types of adolescents. *Psychoanalytic Quarterly*, 1938, *7*, 243–253.

Blasi, A. A developmental approach to responsibility training (Doctoral dissertation, Washington University, 1971). *Dissertation Abstracts International*, 1971, *32*, 1233B.

Blasi, A. Personal responsibility and ego development. In R. deCharms, *Enhancing motivation: Change in the classroom.* New York: Irvington, 1976.

Block, H. J. Conceptions of sex-role: Some cross-cultural and longitudinal perspectives. *American Psychologist*, 1973, *28*, 512–526.

Block, J. *The Q-sort method in personality assessment and psychiatric research.* Springfield, Ill.: Charles C Thomas, 1961.

Block, J. *Lives through time.* Berkeley: Bancroft, 1971.

Block, J. Some enduring and consequential structure of personality. In A. L. Robin (Ed.), *Further explorations in personality.* New York: Wiley, 1981.

Block, J., & Block, J. H. *The developmental continuity of ego control and ego resiliency: Some accomplishments.* Paper presented at the organization of development and the problem of continuity in adaptation symposium at the meeting of the Society for Research in Child Development, New Orleans, March 1977.

Block, J. H., & Block, J. The role of ego control and ego resiliency in the organization of behavior. In W. A. Collins (Ed.), *Minnesota Symposia on Child Psychology* (Vol. 13). New York: Erlbaum, 1979.

Bonneville, L. P. *The relation of role playing and personal characteristics to ego development.* Unpublished doctoral dissertation, Washington University, 1978.

Bourne, E. The state of research on ego identity: A review and appraisal. (Part I). *Journal of Youth and Adolescence*, 1978, *7*(3), 223–254. (a)

Bourne, E. The state of research on ego identity: A review and appraisal. (Part II). *Journal of Youth and Adolescence*, 1978, *7*(4), 371–372. (b)

Broderick, C. B., & Pulliam-Krager, H. Family process and child outcomes. In W. Burr, R. Hill, F. I. Nye, & I. L. Reiss (Eds.), *Theories about the family* (Vol. 1). New York: Macmillan, 1979.

Candee, D. Ego developmental aspects of new left ideology. *Journal of Personality and Social Psychology*, 1974, *30*, 620–630.

Constantinople, A. An Erikson measure of personality development in college students. *Developmental Psychology*, 1969, *1*, 357–372.

Cooper, C. R., Grotevant, H. D., & Condon, S. M. Methodological challenges of selectivity in family interaction: Assessing temporal patterns of individuation. *Journal of Marriage and the Family*, 1982, *44*, 749–754.

Cooper, C. R., Grotevant, H. D., Moore, M. S., & Condon, S. M. Predicting adolescent role taking and identity exploration from family communication patterns: A comparison of one- and two-child families. In T. Falbo (Ed.), *The single-child family.* New York: Guilford Press, 1983.

Cox, N. Prior help, ego development, and helping behavior. *Child Development*, 1974, *45*, 594–603.

Deutsch, H. *Psychology of women* (Vol. 1). New York: Grune & Stratton, 1944.

Doane, J., West, K., Goldstein, M., Rodnick, E., & Jones, J. Parental communication deviance and affective style. *Archives of General Psychiatry*, 1981, *38*, 679–685.

Donovan, E., & Adelson, J. *The adolescent experience.* New York: Wiley, 1966.

Donovan, J. M. Identity status and interpersonal style. *Journal of Youth and Adolescence*, 1975, *4*, 34–55.

Elkind, D. Egocentricism in adolescence. *Child Development*, 1967, *38* 1025–1034.

Engel, M. The stability of the self-concept in adolescence. *Journal of Abnormal Psychology*, 1959, *58*, 74–83.

Epstein, S. The self-concept revisited or a theory of a theory. *American Psychologist*, 1973, *28*, 405–416.

Erikson, E. H. Growth and crises of the "healthy personality." In M. J. E. Senn (Ed.), *Symposium on the healthy personality.* New York: Josiah Macy Jr. Foundation, 1950.

Erikson, E. H. The problem of ego identity. *Journal of the American Psychoanalytic Association*, 1956, *4*, 56–122.

Erikson, E. H. Identity and the life cycle. *Psychological Issues* (Vol. 1, No. 1). New York: International Universities Press, 1959.

Erikson, E. H. *Childhood and society.* New York: Norton, 1963.

Erikson, E. H. *Identity, youth and crisis.* New York: Norton, 1968.

Ferreira, A.J., & Winter, W. D. Decision-making in normal and abnormal two-child families. *Family Process*, 1968, *1*, 17–35. (a)

Ferreira, A. J., & Winter, W. D. Information exchange and silence in normal and abnormal families. *Family Process*, 1968, *1*, 251–276. (b)

Fingarette, H. *The self in transformation.* New York: Basic Books, 1963.

Frank, G. H. The role of the family and the development of psychopathology. *Psychological Bulletin*, 1966, *64*, 191–265.

Frank, S., & Quinlan, D. M. Ego development and female delinquency: A cognitive-developmental approach. *Journal of Abnormal Psychology*, 1976, *85*, 505–510.

Freud, A. *The ego and the mechanisms of defense.* New York: International Universities Press, 1936.

Freud, A. Adolescence. *The psychoanalytic study of the child* (Vol. 13). New York: International Universities Press, 1958.

Gold, S. Relations between levels of ego development and adjustment patterns in adolescents. *Journal of Personality Assessment* 1980, *44*, 630–638.

Goldberger, N. *Breaking the educational lockster: The Simon's Rock experience.* Great Barrington, Mass.: Simon's Rock Early College, 1977.

Goldstein, M., Rodnick, E., Jones, J. E., McPherson, S., & West, K. Familial precursors of schizophrenic spectrum disorders. In L. Wynne, R. L. Cromwell & S. Matthysse. *The nature of schizophrenia.* New York: Wiley, 1978.

Grotevant, H. D., & Cooper, C. R. *Identity formation and role-taking skills in adolescence: An investigation of family structure and family process antecedents.* Final report prepared for National Institute of Child Health and Development, University of Texas at Austin, 1982.

Grotevant, H. D., Thorbecker, W., & Meyer, M. An extension of Marcia's identity status interview into the interpersonal domain. *Journal of Youth and Adolescence*, 1982, *11*(1), 33–47.

Haan, N., Stroud, J., & Holstein, J. Moral and ego stages in relationship to ego processes: A study of "hippie." *Journal of Personality*, 1973, *41*, 596–612.

Haley, J. Experiment with an abnormal family: Testing in restricted communication setting. *Archives of General Psychiatry*, 1967, *17*, 53–63. (a)

Haley, J. Speech sequences of normal and abnormal families with two children present. *Family Process*, 1967, *6*, 81–97. (b)

Harakel, C. M. Ego maturity and interpersonal styles: A multivariate study of Loevinger's theory (Doctoral dissertation, Catholic University, 1971). *Dissertation Abstracts International*, 1971, *32*, 1190B.

Harter, S. Developmental perspectives on the self system. In M. Hetherington (Ed.). *Social development: Carmichael's Manual of Child Psychology*, New York: Wiley, 1984.

Hartmann, H. *Ego psychology and the problem of adaptation*. New York: International Universities Press, 1958.

Harvey, O. J., Hunt, D. E., & Schroeder, H. *Conceptual systems and personality organization*. New York: Wiley, 1961.

Hauser, S. T. *Black and white identity formation*. New York: Wiley, 1971.

Hauser, S. T. Loevinger's model and measure of ego development: A critical review. *Psychology Bulletin*, 1976, *33*(5), 928–955. (a)

Hauser, S. T. Self-image complexity and identity formation in adolescence longitudinal studies. *Journal of Youth and Adolescence*, 1976, *5* 161–177. (b)

Hauser, S. T. The content and structure of adolescent self-images: Longitudinal studies. *Archives of General Psychiatry*, 1976, *33*, 27–32. (c)

Hauser, S. T., & Shapiro, R. Dimensions of adolescent self-images. *Journal of Youth and Adolescence*, 1972, *1*, 339–353.

Hauser, S. T., Pollets, D., Turner, B., Jacobson, A., Powers, S., & Noam, G. Ego development and self-esteem in diabetic youth. *Diabetes Care*, 1979, *2*, 465–471.

Hauser, S. T., Jacobson, A., Powers, S., Schwartz, J., & Noam, G. Family interactions and ego development in diabetic adolescents. In Z. Laron and A. Galatzer (Eds.), *Psychosocial aspects of diabetes in children and adolescents*. Basel: Karger, 1982. (a)

Hauser, S. T., Jacobson, A., Naom, G., & Powers, S. Family influences upon adolescent ego development. *Scientific Proceedings, American Psychiatric Association, 135th Annual Meeting*, 1982. (b)

Hauser, S. T., Jacobson, A., Noam, G., & Powers, S. Ego Development and Self-Image Complexity. *Archives of General Psychiatry*, 1983, *40*, 325–332.

Hauser, S. T., Powers, S. I., Noam, G., Jacobson, A., Weiss, B., & Follansbee, D. Familial contexts of adolescent ego development. *Child Development*, 1984, *55*, 195–213.

Hetherington, E. M., Stouwie, R., & Ridberg, E. Patterns of family interaction and child rearing attitudes related to three dimensions of juvenile delinquency. *Journal of Abnormal Psychology*, 1971, *78*, 160–176.

Hodgson, J. W., & Fischer, J. L. Sex differences in identity and intimacy development in college youth. *Journal of Youth and Adolescence*, 1979, *8*, 32–50.

Holt, R. R. Review of "Measuring ego development. Volumes I and II." *Journal of Nervous and Mental Diseases*, 1974, *158*, 310–316.

Hoppe, C. F. Ego development and conformity behaviors (Doctoral dissertation, Washington University, 1972). *Dissertation Abstracts International*, 1972, *33*, 6060B.

Jacob, T. Patterns of family conflict and dominance as a function of child age and social class. *Developmental Psychology*, 1974, *10*, 1–12.

Jacob, T. Family interaction in disturbed and normal families: A methodological and substantive review. *Psychological Bulletin*, 1975, *82*, 33–65.

Jones, H. E. The California adolescent growth study. *Journal of Education Research*, 1938, *31*, 561–567.

Josselson, R. Psychodynamic aspects of identity formation in college women. *Journal of Youth and Adolescence*, 1973, *2*, 3–52.

Josselson, R., Greenberger, E., & McConochie, D. Phenomenological aspects of psychosocial maturity in adolescence, Part I: Boys. *Journal of Youth and Adolescence*, 1977, *6*, 25–56. (a)

Josselson, R., Greenberger, E., & McConochie, D. Phenomenological aspects of psychosocial maturity in adolescence, Part II: Girls. *Journal of Youth and Adolescence*, 1977, *6*, 145–167. (b)

Josselson, R. Ego development in adolescence. In J. Adelson (Ed.), *Handbook of adolescent psychology*. New York: Wiley, 1980.

Josselyn, I. M. *The adolescent and his world*. New York: Family Services Association of America, 1952.

Kelly, G. A. *The psychology of personal contructs*. New York: Norton, 1955.

Kinsler, P. *Ego identity status and intimacy*. Unpublished doctoral dissertation, State University of New York, 1972.

Kohlberg, L. Development of moral character and moral ideology. In M. Hoffman & L. Hoffman (Eds.), *Review of child development research* (Vol. I). New York: Russel Sage, 1964.

Lambert, H. V. *A comparison of Jane Loevinger's theory of ego development and Lawrence Kohlberg's theory of moral development*. Unpublished doctoral dissertation, University of Chicago, 1972.

Lewis, J. M., Beavers, W. R., Gossett, J. T., & Phillips, V. A. *No single thread*. New York: Brunner/Mazel, 1976.

Lidz, T., Cornelison, A. R., Fleck, S., & Terry, D. The interfamilial environment of schizophrenic patients: II: Marital schism and marital skew. *American Journal of Psychiatry*, 1957, *114*, 241–248.

Loevinger, J. The meaning and measurement of ego development. *American Psychologist*, 1966, *21*, 195–206.

Loevinger, J. *Ego development: Conceptions and theories*. San Francisco: Jossey-Bass, 1976.

Loevinger, J. *Scientific ways in the study of ego development*. Worchester, Mass.: Clark University Press, 1979.

Loevinger, J., & Wessler, R. *Measuring ego development* (Vol. I). San Francisco: Jossey-Bass, 1970.

Loevinger, J., Wessler, R., & Redmore, C. *Measuring ego development* (Vol. 2). San Francisco: Jossey-Bass, 1970.

Lynch, M. D. Self-concept development in childhood. In M. D. Lynch, A. A. Norem-Hebeisen, & K. Gergen (Eds.), *Self concept: Advances in theory and research*. Cambridge, Mass.: Ballinger, 1981.

MacFarlane, J. W. Studies in child guidance I. Methodology of data collection and organization. *Monographs of the Society for Research in Child Development*, 1938, *3* (6, Whole No. 19).

Marcia, J. E. *Scoring manual for ego identity status*. Unpublished manuscript, Simon Fraser University, British Columbia, Canada, 1964.

Marcia, J. E. Development and validation of ego development status. *Journal of Personality Social Psychology*, 1966, *3*, 157–166.

Marcia, J. E. Ego identity status: Relationship to change in self-esteem, "general maladjustment," and authoritarianism. *Journal of Personality*, 1967, *35*, 113–133.

Marcia, J. E. *Studies in Ego Identity*. Burnuby, British Columbia: Simon Fraser University, 1976. (a)

Marcia, J. E. Identity six years after: A follow-up study. *Journal of Youth and Adolescence*, 1976, *15*, 145–150. (b)

Marcia, J. E. Identity in adolescence. In J. Adelson (Ed.), *Handbook of adolescent psychology*, New York: Wiley, 1980.

Marcia, J. E., & Friedman, M. L. Ego identity status of college women. *Journal of Personality*, 1970, *38*, 249–63.

Marcus, D., Offer, D., & Platt, S. A clinical approach to the understanding of normal and pathologic adolescence. *Archives of General Psychiatry*, 1966, *15*, 569–576.

Markus, H. Self-schemata and processing information about the self. *Journal of Personality and Social Psychology*, 1977, *35*, 63–78.

Matteson, D. R. *Adolescence today: Sex roles and the search for identity.* Homewood, Ill.: Dorsey, 1975.

Matteson, D. R. Exploration and commitment: Sex differences and methodological problems in the use of identity status categories. *Journal of Youth and Adolescence*, 1977, 353–374.

McPherson, S. Communication of intents among parents and their disturbed adolescent child. *Journal of Abnormal Psychology*, 1970, *76*, 98–105.

McPherson, S., Goldstein, M., & Rodnick, E. Who listens? Who communicates? How? *Archives of General Psychiatry*, 1973, *28*, 393–399.

Mead, D. E., & Campbell, S. S. Decision-making and interaction by families with and without a drug abusing child. *Family Processes*, 1972, *11*, 487–498.

Minuchin, S. *Families and family therapy*, Cambridge: Harvard University Press, 1974.

Minuchin, S., Rosman, B., & Baker, L. *Psychosomatic families*. Cambridge: Harvard University Press, 1978.

Mishler, E. G., & Waxler, N. E. *Interaction in families*. New York: Wiley, 1968.

Munro, G. & Adams, G. R. Ego identity formation in college students and working youth. *Developmental Psychology*, 1977, *13(5)*, 523–524.

Noam, G. G., Kohlberg, L. & Snarey, J. Steps toward a model of the self. In B. Lee & G. Noam (Eds.), Developmental approaches to the self. New York: Plenum Press, 1982.

Offer, D. *The psychological world of the teenager.* New York: Basic Books, 1969.

Offer, D., & Offer, J. B. *From teenage to young manhood.* New York: Basic Books, 1975.

Orlofsky, J. L. Identity formation. Achievement, and fear of success in college men and women. *Journal of Youth and Adolescence*, 1978, *7*, 49–62.

Orlofsky, J. L., Marcia, J. F., & Lesser, L. M. Ego identity status and the intimacy vs. isolation crisis of young adulthood. *Journal of Personality and Social Psychology*, 1973, *27*, 211–219.

Piaget, J. *The moral judgment of the child.* London: Kegan Paul, 1932.

Raphael, D., & Xelowski, H. G. Identity status in high school students: A critique and a revised paradigm. *Journal of Youth and Adolescence.* 1980, *9(5)*, 383–389.

Redmore, C. D., & Loevinger, J. Ego development in adolescence: Longitudinal studies. *Journal of Youth and Adolescence*, 1979 *8(1)*, 1–20.

Redmore, C. D., & Waldman, K. Reliability of a sentence completion measure of ego development. *Journal of Personality Assessment*, 1975, *39*, 236–243.

Riskin, J., & Faunce, E. E. An evaluative review of family interaction research. *Family Process*, 1972, *11*, 365–456.

St. Clair, S., & Day, H. D. Ego identity status and values among high school females. *Journal of Youth and Adolescence*, 1979, *8(3)*, 317–319.

Sarbin, T. R. A preface to a psychological analysis of the self. *Psychological Review*, 1962, *59*, 11–22.

Schain, W. S. Psychodynamic factors affecting women's occupational choice: Parent-child relations, expressed needs and level of ego development (Doctoral dissertation, George Washington University, 1974). *Dissertation Abstracts International*, 1974, *35*, 1991A.

Schenkel, S., & Marcia, J. Attitudes toward premarital intercourse in determining identity status in college women. *Journal of Personality*, 1972, *40(3)*, 473–489.

Sheridan, S. J. W. Level of moral reasoning and ego development as factors in predicted vocational success with the mentally retarded (Doctoral dissertation, University of Houston, 1974). *Dissertation Abstracts International*, 1974, 3395B.

Singer, M., & Wynne, L. Thought disorder and family relations of schizophrenics: IV: Results and implications. *Archives of General Psychiatry*, 1965, *12*, 201.

Stabenau, J. R., Tupin, J., Werner, M, & Pollin, W. D. A comparative study of families of schizophrenics, delinquents and normals. *Psychiatry*, 1965, *28*, 45–59.

Stephenson, W. *The study of behavior*. Chicago: University of Chicago Press, 1953.

Stierlin, H. *Separating parents and adolescents*. New York: Quadrangel, 1974.

Stone, G. C., Cohen, F., & Adler, N. (Eds.). *Health psychology: A handbook*. San Francisco: Jossey-Bass, 1979.

Strauss, A. L. *Mirrors and masks: The search for identity*. New York: MacMillan, 1959.

Strodtbeck, F. L. Husband–wife interaction and revealed differences. *American Sociological Review*, 1958, *16*, 468–473.

Sullivan, C., Grant, M. Q., & Grant, J. D. The developmental of interpersonal maturity; Applications to delinquency. *Psychiatry*, 1957, *20*, 373–385.

Sullivan, H. S. *The interpersonal theory of psychiatry*. New York: Norton, 1953.

Sullivan, P. J. A curriculum for stimulating moral reasoning and ego development in adolescents. (Doctoral dissertation, Boston University, 1975). *Dissertation Abstracts International*, 1975, 1320A.

Toder, N., & Marcia, J. E. Ego identity status and response to conformity pressure in college women. *Journal of Personality and Social Psychology*, 1983, *26*, 287–294.

Vincent, L., & Vincent, K. Ego development and psychopathology. *Psychological Report*, 1979, *44*, 408–410.

Waterman, A. S. Identity development from adolescence to adulthood: An extension of theory and a review of research. *Development Psychology*, 1982, *18(3)*, 341–358.

Waterman, A. S., & Waterman, C. K. Relationship between freshman ego identity status and subsequent behavior: A test of the predictive validity of Marcia's categorization system for identity status. *Developmental Psychology*, 1972, *6(1)*, 179.

Waterman, C. K., Bruebel, M. E., & Waterman, A. J. The relationship between resolution of the identity crisis and outcomes of previous psychosocial crises. *Proceedings of the 78th Annual Convention of the American Psychological Association*, 1970, *5*, 467–468. (Summary)

Waugh, M. H., & McCaulley, M. H. Relation of level of ego development to type and severity of psychopathology. *Journal of Consulting and Clinical Psychology*, 1981, *49(2)*, 295–296.

Weiner, I. B. Psychopathology in adolescence. In J. Adelson (Ed.), *Handbook of adolescent pscyhology*. New York: Wiley, 1980.

Wheelis, W. *The quest for identity*. New York: Norton, 1958.

White, K. M., & Speisman, J. C. Adolescence. In F. Rebelsky & L. Dorman (Eds.), *Life span human development series*. Monterey, Calif.: Brooks/Cole, 1977.

Wylie, R. *The self-concept: Volume 1. A review of methodological considerations and measuring instruments* (Rev. ed.). Lincoln: University of Nebraska Press, 1974.

Wylie, R. *The self-concept: Volume 2. Theory and research on selected topics* (Rev. ed). Lincoln: University of Nebraska Press, 1978.

Wynne, L. C. & Singer, M. Thought disorder and family relations of schizophrenics: II: A classification of forms of thinking. *Archives of General Psychiatry*, 1963, *9*, 199–206.

Wynne, L. C., Ryckoff, I. M., Day, J., & Hirsch, S. I. Pseudomutuality in the family relations of schizophrenics. *Psychiatry*, 1958, *21*, 205–220.

Author Index

269

274 AUTHOR INDEX

Subject Index